Mathematical Underpinnings of Analytics

MATHEMATICAL UNDERPINNINGS OF ANALYTICS

Theory and Applications

PETER GRINDROD CBE

Mathematical Institute
University of Oxford

OXFORD
UNIVERSITY PRESS

OXFORD
UNIVERSITY PRESS

Great Clarendon Street, Oxford, OX2 6DP,
United Kingdom

Oxford University Press is a department of the University of Oxford.
It furthers the University's objective of excellence in research, scholarship,
and education by publishing worldwide. Oxford is a registered trade mark of
Oxford University Press in the UK and in certain other countries

First Edition published in 2015

Impression: 1

Published in the United States of America by Oxford University Press
198 Madison Avenue, New York, NY 10016, United States of America

British Library Cataloguing in Publication Data

Data available

Library of Congress Control Number: 2014948658

ISBN 978-0-19-872509-1

Printed and bound by
CPI Group (UK) Ltd, Croydon, CR0 4YY

To Dora, Tom, Chris, Jumbly, and Sophie.

He who would learn to fly one day must first learn to stand and walk and run and climb and dance; one cannot fly into flying.
Friedrich Nietzsche

I think I'm constantly in a state of adjustment.
Patti Smith

PREFACE

S tarting out from a mathematical standpoint, this book introduces a wide range of the concepts, methods and applications that are current within analytics. As I set out in the Introduction, the topics of analytics and data science within customer-facing sectors are really the practical interface of a much larger field of study that could be termed the 'mathematics of human behaviour'. Accordingly, there are two introductory accounts given there: one explaining the commercial and technological drivers of analytics and its value within the digital economy; the other discussing the evolving face of mathematical modelling, explaining why this field is a natural next step for the applied mathematical sciences. I hope that this book will serve as a text for students wishing to develop their interests in analytics, from both theoretical and practical perspectives, as well as for early career professionals within commercial analytics teams, as a source of background experience, ideas and, benchmarks. My wish is that more and more mathematical sciences graduates will become influential leaders within fields of commercial analytics.

What should be clear is that mathematics is highly differentiating in producing new methods and algorithms, not least in dealing with uncertainties. We might often have a lot of data, but it may contain a lot of errors. We should avoid methods that simply grind out metrics telling us what is there. With mathematical models based on rigorous foundations we may gain insights and ask, 'What might be there?', 'What does it mean, what can we do now?' or even, 'What have we not observed?'

The theory and examples discussed here reflect my own interests in social networks, peer-to-peer communication, demand behaviour, purchasing behaviour, customer behaviour, social norms, and lifestyle changes. I have selected ideas and applications that are useful to customer-facing businesses, such as retailers and consumer goods manufacturers, e-commerce companies, mobile network operators, digital media and marketing companies, energy distributers, and finance, insurance, health, and leisure providers. Necessarily these are very modern applications and the selection of material here is subjective and personal to me. I make no attempt to review, nor claim to be exhaustive. Data sets associated with the book can be freely accessed online <http://www.oup.co.uk/companion/analytics>.

At the end of each chapter I have included some personal views on matters relating to the theory and the wider commercial and academic contexts of analytics. This includes advice on a variety of topics that some readers may find useful.

I would like to acknowledge the huge amount of assistance, competitive challenge, and encouragement that my colleagues at Numbercraft, Bloom Agency, Counting Lab, Cignifi, and Quintessa gave me within my research over many years working within diverse areas of analytics, quantitative assessment, forecasting and inference, and modelling under very different types of uncertainty. The time spent working with these teams has been both stimulating and fun.

For a number of years I have collaborated very happily and productively with Des J. Higham. I have gained a lot from him, and benefitted from his enthusiasm, good council, creativity, and sense of humour. He has made me strive to work harder.

The material presented here relied on the efforts of my many colleagues and the co-authors of various parts of the underlying material, especially Andy Briant, Robert Brown, Billiejoe Charlton, Sam Clarke, Alex Craven, Ernesto Estrada, Jon Flitton, Danica Vukadinovic Greetham, Rebecca Gower, Simon Grindrod, Stephen Haben, Chrystalla Hadjipavlou, Richard Hibbert, Gabriela Kalna, Guy Keeling, Milla Kibble, Sharon Kirkham, Dan Klinger, Peter Laflin, James Laurie, Tamsin Lee, Mark Parsons, Nick Rafferty, Alain Reuter, Doug Saddy, Colin Singleton, Alastair Spence, Zhivko Stoyanov, Andrew 'Lol' Tallack, Chris Tandy, Keith Vass, Jonathan Ward and David 'Muddy' Waters.

I would like to thank Clive Bowman for his patient advice and many conversations with me about discrimination, Bayes factors, and much more; and Simon Chandler-Wilde who encouraged me when I returned to academia after years away.

I am indebted to my friend Robert 'Roy' C. Brown who has both tolerated and supported my way of working (and has removed many errors from my thinking), and to Tom Grindrod who has corrected and improved drafts of this manuscript.

Some of the research described here has been supported by the UK's Engineering and Physical Sciences Research Council, through the funding for the Horizon Digital Economy Hub and the Mathematical Underpinnings of The Digital Economy. This allowed me to develop theories and some new relationships with exploiters across a number of customer facing sectors.

Finally, I would like to thank my friends and colleagues in the Mathematical Institute at the University of Oxford, and within the wider maths and analytics communities, for encouraging me to burst into print.

Peter Grindrod
Oxford, April 2014.

CONTENTS

Introduction: The Underpinnings of Analytics

Aspirations

In almost every sector of commercial and public endeavour there has been, or there is about to be, a data deluge. The innovation and exploitation, and also the hype, are driven by (a) the availability of data from emerging and converging digital platforms; (b) the increased amount of online and off-line traffic, data collection, and surveillance; (c) the commercial imperatives to create greater value from existing customers and distilled knowledge; and (d) growing open data initiatives. Companies have become more aware of their own data resources, and see the future exploitation of these resources as a strategic path to growth.

When the data are very large, or continuously arrive very rapidly, doing anything smart is heavy lifting, and doing anything smart in real time is a challenge. Data have volume and velocity. Given enough processing time and capacity anybody might achieve anything. Yet companies and institutes have, in many cases, already invested their money in their infrastructure, architecture, enterprise, platforms, and access. Now they need to see some value return and value growth.

Mathematical Underpinnings of Analytics. First Edition. Peter Grindrod.
© Peter Grindrod 2015. Published in 2015 by Oxford University Press.

However, the data itself are merely the raw material. It is nothing without *analytics*[1]: the concepts, methods, and practices the can conjure valuable and actionable insights and radical knowledge from the large volumes of data. Such smart analytics goes well beyond 'dice and slice', alerts, reporting and, querying, and it is far beyond the provision of infrastructure, computing architectures, and data handling resources. The latter are all necessary but are simply not sufficient for success. It is the analytics that will provide distinctiveness and unique capabilities, and allow us to see 'further than others'[2]. Such analytics needs to be founded on rigorous mathematical concepts, ideas, and methods that may be deployed so as to underpin and innovate new concepts, methods, and algorithms. In turn, this could inspire new products and services and expand what companies and public institutes could achieve. It could even give rise to new business models. Companies can and will differentiate themselves on their exploitation of such analytics, and there clearly is an opportunity for them to obtain more insight and business growth from their data resources, going beyond current operations.

As different forms of data become evermore pervasive and more available, so the next generation of businesses and services within significant sectors (including commercial services, digital media, communications, domestic energy, security, environment, marketing, targeting, customer relationship management) and engagement across almost all public sectors will be developed by those companies, small and large, best placed to innovate. Our digital platforms will evolve rapidly year on year, converging and becoming pervasive with 24/7 operations, and data will become more open, as a currency or commodity.

Yet what is modern data science? What will data scientists need to achieve? This is rather a crowded space. Other voices and opinion-formers may argue from their own experience that it must subsume some or all of the following issues and activities.

Foundations: Novel Theoretical Models, Computational Models, Data and Information, Data Standards.

Infrastructure: Cloud/Grid/Stream, High Performance/Parallel Computing, Platforms, Autonomic Computing, Cyber-infrastructure, System Architectures, Design and Deployment, Energy-efficient Computing, Models and

---

[1] 'Competing on analytics' championed by Davenport in the Harvard Business Review [33] and elsewhere.
[2] 'If I have seen further than others, it is by standing upon the shoulders of giants.' Isaac Newton

Environments for Cluster, Software Techniques and Architectures, Open Platforms, New Programming Models beyond Hadoop/MapReduce, Software Systems.

Management: Advanced Database and Web Applications, Novel Data Model, Databases for Emerging Hardware, Data Preservation, Data Provenance, Interfaces to Database Systems and Analytics Software Systems, Data Protection, Integrity and Privacy Standards and Policies, Information Integration and Heterogeneous and Multi-structured Data Integration, Data management for Mobile and Pervasive Computing, Data Management in the Social Web, Crowdsourcing, Spatiotemporal and Stream Data Management, Scientific Data Management, Workflow Optimization.

Database Management Challenges: Architecture, Storage, User Interfaces.

Security and Privacy: Intrusion Detection for Gigabit Networks, Anomaly and APT Detection in Very Large Scale Systems, High Performance Cryptography, Visualizing Large Scale Security Data, Threat Detection using Big Data Analytics, Privacy Threats of Big Data, Privacy Preserving Big Data Collection/Analytics, HCI Challenges for Big Data Security and Privacy, User Studies for any of the above, Sociological Aspects of Big Data Privacy.

Search and Mining: Social Web Search and Mining, Web Search, Algorithms and Systems for Big Data Search, Distributed and Peer-to-peer Search, Big Data Search Architectures, Scalability and Efficiency, Data Acquisition, Integration, Cleaning, and Best Practices, Visualization Analytics for Big Data, Computational Modelling and Data Integration, Cloud/Grid/Stream Data Mining-Big Velocity Data, Link and Graph Mining, Semantic-based Data Mining and Data Pre-processing, Mobility, Multimedia and Multi-structured Data.

This book is about none of these things, but any successful analytics strategies, operations, and methods will surely rely on elements of them. In this book we shall focus on the mathematics that underpins world-class analytics, and thus creates some differentiation and value.

It is arguable that game changing capability must come from creative applications of analytics to novel data sources, since access to digital platforms and computational power (in-house or in-the-cloud) is now ubiquitous. In any case, most companies have made (or will make) decisions on a long-term basis about their operations (computer resources, data architectures, enterprise systems, and so on), and they, together with their suppliers, should be evermore focussed on getting a return from those investments and plans. Hence their need for creative analytics and novel theoretical concepts that may in time become applicable, deliver actionable insights, and, thus, some agility and an ability to act.

Big data is presently on an trajectory resembling the Gartner hype curve. It is confusingly disorganized, because it should really be split into at least four different challenges within data science. This way we prevent groups of people contributing very different things, with distinct aims, all claiming to be doing the essential element of vast data science.

1 There is vast data in the sense that we know a little bit about a very large number of things/objects/people. For example, in customer-facing industries such as mobile telcos, retailers, ISPs, retail banks, smart metering in energy, digital marketing, and social media, etc. We call these customer- or consumer-facing applications, where the data is often proprietary.
2 There is vast data in the sense that it is constantly arising (has velocity) and so real-time analytics are required. For example, analysis of social media, or monitoring peer-to-peer conversations, emails, messaging, financial market data, or real time response to e-commerce, etc. Again, this data is often proprietary.
3 There is vast data in the sense that we know a very large amount about a relatively small number complex objects, such as images for transmission and compression, scientific applications from crystallography, proteomics, fMRI scans, or spectroscopy, etc. We call these scientific or engineering applications, where the high throughput of novel scientific equipment spews out extremely high resolution data.
4 There is vast data in the sense that we know medium amounts about each of a number of distinct types of data objects belonging to individuals, which we will join or match together. For example, by joining omic data and clinical data (unstructured). We call these multi-source applications. The data is big through the consolidation of the distinct data resources, for example, in health initiatives, public sector, security, etc.

Primarily we shall be dealing with the first two of these challenges, and less so with the issues raised by the latter, since this is where commercial interests lie and we are perhaps working over proprietary data. Occasionally we shall digress into the other fields where this is helpful to do so and there is some common cause.

This book is for mathematicians, quantitative scientists, and engineers who have had some exposure to mathematics during a first degree. It is written for final year undergraduate and first year postgraduate students. Hopefully it will encourage an experimental approach, yet provide a coherent theoretical framework by founding the concepts firmly within mathematics.

The key challenge for those of us researching, working with, and exploiting analytics is to produce insights that are data-driven, are usually hidden at first sight, and reflect some knowledge that is novel or private to the data owner, and thus advantageous. Even seasoned professionals are not able to guess what the consequent best action to take is, or which are the best opportunities to exploit. They seek business growth within a competitive environment. So our outputs should be quantitative (what might happen?) and actionable (what are the priorities, most likely scenarios, and possible tactics?).

A common problem is that analytics teams often do things just because they can. They make the mistake of producing multiple outputs covering all possible applications of a given method, without priority. A key word here is *distillation*; a lovely word. We must distil the outputs from the data. Many of us have sat through exhaustive presentations where diligent analysts have turned some handle and converted a kilogram of data into a kilogram of powerpoint, and they expected our gratitude. Decisions have to be both evidence-based and justified. Yet they should be 'smart' because they are data-driven.

'This insight is data-driven we did not hypothesize this, we found it', whereas the IT team's favourite hypothesis-driven analytics is often merely dice and slice, so it, 'just reports what they think that you need to know.' 'There are no more compelling options available', and so, 'one can rest assured that there is no better hypothesis or option that one could have checked'. 'These are options/tactics listed in priority order', so these investments can be ranked and justified.

Access to a range of mathematical ideas and methods is essential for success. Within commercial environments, when challenged by client-companies or business stakeholders, it is often necessary to get a team of analytics folk together to argue over alternative approaches and to sense-check the usefulness of the outputs that different approaches might provide. Two heads are always better than one when planning analytics. We need both effectiveness and efficiency, and these are not to be confused.

As the data gets very large, or it arrives at a high rate, computational efficiencies become very important. The good news is that computational power and resource are ever increasing. The bad news is that the data deluge never ceases. So our key asset must be an understanding of what various methods may and may not achieve. As mathematicians we habitually abstract and translate analytics concepts and methods into new domains to solve new problems, but we cannot do so innovatively and securely without understanding the rigorous *mathematical underpinnings of analytics*. This is why considerations such as those set out in this book are foundations for a career in analytics and not simply 'how to apply' notes for a set of pre-cooked algorithms.

In each chapter we shall introduce a mathematical framework and present some theory, before launching into illustrative applications that set out the strengths and limitations of the ideas. The reader will also be provided with links to data resources and will be encouraged to experiment for themselves. The reader should be familiar with at least some the material set out in the Appendix. If not, then please start there or consult it as terms comes up within the main text. This has been provided both so that readers can revisit and explore basic considerations, and to make this book more self-contained, harmonizing the notation employed throughout. Each chapter will have some suggestions for project case studies and further work, and wherever possible an appropriate data set has been made available or is referenced directly to download and explore.

Nuts and Bolts

Nowadays, mathematics is evolving rapidly with respect to its external fields of application, and thus impact, while internally it is grinding on. Against this backdrop we all travel on our own journey. The fundamentals of analysis, algebra, geometry, and calculus provide us with rigorous frameworks, concepts, vocabulary, methods, and notation. Indeed, the last of these can often be the most important with regard to communication, abstraction, and simplification. The applications (the external ramifications of mathematics) are changing though, and changing at a faster pace than many professionals may care to concede.

For many hundreds of years since the Renaissance, mathematics was the key to an understanding of the natural—mostly physical—sciences. It reached its zenith in providing insights into physical phenomena on the scales of time and space most easily explorable, and thus most easily comprehensible and exploitable, by mankind. The observations from celestial mechanics, previously a mystery of the ancients, gave way to the understating of gravity and many other natural physical phenomena. Broadly speaking, all such applications have a foundation of conservation laws at small scales: conversation of mass, momenta, energy. They are integrals for the fundamental equations of motion, reflecting symmetries at work. (We shall seek our own symmetries later in this text.) They usually give rise to continuum models for linear and nonlinear rates of change which have become phenomenally successful, expressed in terms of partial differential equations.

Just over one hundred years ago, the science of the very large and the very small (scales, velocities, forces) provided a novel set of challenges: not just in explicating newly observable phenomena, but in providing theoretical support to as yet unobserved possibilities. Physics thus shifted from justification and explanation to inference and hypothesis creation. Statistical mechanics and stochastic fluctuations provided further challenges in passing from the small to the large scales. Meanwhile dynamical systems, even of deterministic systems, emerged as a major source of uncertainty, sensitive dependence, unpredictability, structure, form, and pattern.

By the middle of the twentieth century, in the UK at least, the continuum mechanics paradigm was dominant in applied mathematics. From the Navier-Stokes equations in fluid mechanics to Maxwell's equations in electromagnetics (and their plethora of approximations and simplifications), there was both willing momentum from within the science and a strong desire from within the applications (industry, military, communications). Weak solutions (shocks), boundary layers, and moving boundary problems of all kinds stretched the knowledge further.

Then the thrilling emergence of modern computing changed the way that mathematicians could work forever. The birth of numerical analysis not only responded to a desire to calculate, simulate, and predict, but also yielded novel topics in its own right: optimization in all its forms; finite differences and elements; and glorious numerical linear algebra. The growth in computational resources also heralded the return of some older mathematical ideas that had remained largely parked and impotent for almost two hundred years, the most important of which was Bayesian reasoning. The previous difficulties in its application were merely practical. Only simple or contrived examples that were reliant on useful trickery, such a conjugate priors, were amenable to analysis: otherwise posterior distributions became easy to think about yet largely inaccessible. The consequence was that probability theory lost its meaning. The rise of frequentist 'cooking' methods was welcomed by many practitioners within social and scientific applications, yet was fundamentally and intellectually bankrupt at birth. Computational resource reversed this calumny, not least with sampling methods. Today, a week or two spent reading Jaynes' book [83] can be a life-changing experience.

From the 1970s onwards, these, and other, mathematical ideas found two new outlets: the mathematics of finance and risk, and the mathematics of the life sciences. These fields provided two distinct types of challenge: the former being the need to qualify uncertainty, risk, and, thus, value; the latter being about form and function (both normal and aberrant). By the end of the 1980s almost all universities had undergraduate courses on some elements of mathematical finance and mathematical biology, not least because these fields are large employers of graduates, and also because they were both about to explode. Ultimately one was driven by the global information and communication avalanche enabling electronic trading, and the other was driven by the informatics from high throughput genomics. A data deluge drove these subjects.

Though disruptive to the old school, these transitions were very comfortable for most mathematicians, relying largely on continuum concepts and models. The words changed but the song remained the same.

However, somewhere in the past forty years, a schism crept in. There is nowadays a denominational division between methods-based applied mathematicians, whom we shall call pragmatists, and the theory-based applied mathematicians, whom we shall call theorists. The pragmatists prize methods, learning through doing, and analytical approaches that exploit the particular structure of equations and problems. They have a great eye for problems, they zoom in and chase details, and their applied mathematics is a plumbers' bag of exotic tools. Meanwhile the theorists integrate the applications within a wider phenomenology and they use them to challenge and expand both the theory and the applied activity. They generalize and abstract to ever bigger problem classes.

The biggest losers from a restricted pragmatist's diet are students. They take the short-term gains of being able to answer some standard problems (oft rather special cases, oft learned by rote), but giving up, ignoring, and not challenging themselves with any mention of the larger problems. The normal response to this would be to have a foot in both camps, but this is often not achieved, especially given the polarization of the interests of lecturers.

One should always resist over-reliance on methods and pragmatics. One might spot oneself behaving as a man with a hammer who is obsessively searching for nails. At best, if successful, this leads to a virtuoso's skill. At worst, one's appetite and horizons become restricted and impact, esteem, and influence will forever be limited to a smallish community of similarly-obsessed *cognoscenti*.

The antidote is adventure, the launchpad for which is almost always found at the intersection between some fundamental mathematical idea and the allure of some fresh field of application.

Mathematics is subject to a further malaise called the 'King Herod' principle. Established fields, championed by their participants, seek to weaken and extinguish novel and radical topics in mathematics while they are embryonic or infantile, and before they can grow up and compete for research funding, staffing, and esteem. This is natural in many ways, but it is also destructive and murderous. As the funding situation gets tight in straightened times, all sorts of arguments based on critical mass, retention of key players, and continuity of perceived critical competencies are trotted out to justify the *massacre of innocent* ideas.

It is against these background forces that a new field of applied mathematics has staked its claim: the *mathematics of behavioural analytics*. This was driven by the data deluge from novel digital platforms, and could equally be termed the *mathematics of the digital society*.

Yet even before academic mathematicians became aware of this, the genie was already out of the bottle. Many companies and institutions simply could not wait for the mathematical research community to catch up with the applications. The solutions to disruptive challenges and the novel opportunities so created were simply too valuable. The result was that many of the pioneering ideas in 'analytics' (the mathematical and quantitative analysis of data resources) were produced by small research groups working within industry, and especially within start-up companies. Even the vocabulary was that of business competitiveness, and, for many analytics practitioners like the author, it was essential that analytics was seen as the provider of a competitive edge, and an activity championed in business schools and adopted in boardrooms long before it crossed over into academic research with the mathematical sciences. In particular, Davenport (see [33], for example) was a highly influential figure in arguing for this culture change. The challenges and benefits that were set out in such work, drawn against the backdrop of the explosive growth in the digital society and commerce, certainly made it straightforward to grow analytics practices. The demand for creativity in analytics, and thus the insights to drive highly differentiating and competitive actions/options, is still accelerating.

In turn, this requires both sure-footedness and good decision-making skills of analytics professionals, and these can only come from a firm foundation for models, methods, and algorithms within mathematics and some experience (exemplars). Over the next few years there will be a deluge of analytics taught in universities as part of graduate-level courses and professional development courses on Data Science, Big Data and Analytics, and Discrete Modelling for Commercial Sectors. The mathematical sciences community urgently requires some scholarship and exemplars to define and drive the leading edges of theory and practice. This book is a contribution to that effort.

The birth of any new field of mathematical applications is never entirely straightforward, yet it has happened. And now, after the physical and life sciences, it is the turn of the social sciences to become transformed from being a retrospective and narrative-led theory (telling us what may have happened) into a speculative, insightful, and forward-looking activity (telling us what might happen).

Of course in any new field of mathematical application there is also risk to the mathematicians who are involved. The provision of leadership and scholarship itself requires people to commit to these new fields that, at best, colleagues and others simply may not value, understand, or accept. At worst they may be derided or despised. This should not happen, but we know it does.

Happily one's adherence to the new discipline of *mathematics for behavioural analytics* may be validated every time one speaks to the potential exploiters, especially within service sector companies. For example, those within retail, consumer goods, telecoms, online businesses, energy, finance, betting, leisure, health, IT and communications (including software and services). All of these sectors' operational and research groups will recognize the value of analysis that rises to these digital challenges. They are intrinsic to the future success of our companies, our economy, and our international competitiveness, so we can take comfort from this very strong interest from our potential exploiters. These are companies that work in hugely competitive sectors, that are internationally excellent, and where every decision for investment and activity is tensioned. It can be argued that we are already living through a 'boiling up' phase, prior to the setting out of a formal foundation and scope. Perhaps in ten years, time, the *mathematics of behavioural analytics* will be common place: every mathematics department will be doing some of it. Industry and commerce need it, the government and public regulators will require it, and our students will be attracted towards it. Economic growth, careers, entrepreneurial opportunities, and research challenges will be the drivers.

Necessarily, novel ideas and methods have arrived piecemeal, by trial and error, so it is essential now to avoid the pitfalls of the pragmatists. The comparison to earlier applications of applied mathematics, and the harmful schism with its ignorance of theoretical foundations, is compelling. Let us avoid this from today. We shall deal with discrete mathematics (graphs and networks), probability and inference (forecasting and unsupervised discrimination), optimization (calibration by discrete non-derivative search as well as continuous gradient methods), and system dynamics (stability, response, and structure).

Should not the key skill of the analytics professional be the perspective to see their challenges within the bigger picture? We need a balanced mixture between pragmatics (method—what works?) with the theoretical underpinnings (why and how might anything *work*?).

CHAPTER 1

Similarity, graphs and networks, random matrices, and SVD

S uppose that we are given a set of n objects and a complete list of their $n(n-1)/2$ pairwise comparisons. If this comparison results in a binary indicator then we have an *undirected network* (the objects are vertices of the graph, and the pairwise indicators equal to one represented by the edges). That network has a symmetric adjacency matrix (see Section 1.4). If the $n(n-1)/2$ pairwise comparisons are given in terms of some real measure (such as correlation, for example, or some other types of inner product), then we have a real symmetric similarity matrix. In either case, we may want to reorder the whole set of objects into subsets that are relatively well connected or contain very similar elements, with only small connections or sparse similarities between members of different subsets.

This situation is similar to the concepts of clustering, discussed in Chapter 2, but here it is network-/matrix-based, and we seek something softer than a strict partition. We shall consider generalized clustering methods within a probabilistic setting as a latent (hidden) variable problem. In this chapter, we shall consider how (computational) linear algebra can get us a rather long way through this problem, especially when n is large. We shall do so first from a theoretical perspective and then we shall show how considerations within a field of application actually lead us to reinvent some well-known ideas from matrix factorization.

Mathematical Underpinnings of Analytics. First Edition. Peter Grindrod.
© Peter Grindrod 2015. Published in 2015 by Oxford University Press.

Thus there is a push and a pull, from *linear algebra* and *network theory* (with a helping hand from probability theory), and from *the fields of application*, respectively.

An introduction to Bayesian probability is provided in the Appendix that covers notation and ideas that will be used throughout this book. It may be useful to review some of that material as required, and indeed it can serve as a rapid introduction for those starting from scratch.

Now, let us look at some basic considerations from linear algebra.

1.1 Self-Adjoint Matrices

A self-adjoint matrix, or Hermitian matrix, A, is such that $A = A^*$; the complex conjugate transpose of A. If A is real then this means $A = A^T$. Every self-adjoint matrix is a *normal* matrix, meaning that it commutes with its transpose and is thus diagonalizable. Indeed the spectral theorem says that any real self-adjoint matrix can be diagonalized by a unitary matrix ($U^T A U$ is diagonal, where U is unitary), and that the resulting diagonal matrix has only real entries. This implies that all eigenvalues of a self-adjoint matrix are real, and that it has a full set of linearly independent eigenvectors.

Exercise 1.1

Show that all of the eigenvalues of a self-adjoint matrix, A, are real.

1.2 Non-negative Matrices and the Perron–Frobenius Theorem

Non-negative matrices play a very important role within many theories including graph theory, unsupervised discrimination, mathematical economics, and Markov chain theory. Non-negative symmetric matrices are especially interesting since their spectrum is real. Many courses on linear algebra do not emphasize the properties of non-negative matrices, but within analytics they occur quite often.

Perron–Frobenius Theorem

The Perron–Frobenius theorem gives us a ready way to estimate the spectral radius of non-negative matrices and other things besides. It was proved by Oskar Perron and Georg Frobenius and says that a real square matrix with strictly positive entries has a unique largest real eigenvalue and that the corresponding eigenvector has strictly positive components, and it also asserts a similar statement for certain classes of non-negative matrices, called *irreducible* non-negative matrices.

A non-negative matrix A is irreducible if, for every pair of indices i and j, there exists a natural number k, depending on (i,j), such that $(A^k)_{ij} > 0$. If A is real and $A > 0$ then it is trivially irreducible.

We state the following standard result without proof.

Let A be an irreducible non-negative $n \times n$ matrix with spectral radius $\rho(A) = r > 0$. Then the following statements hold.

(a) r is an eigenvalue of the matrix A, called the Perron–Frobenius eigenvalue.

(b) The Perron–Frobenius eigenvalue r is simple. Both right and left eigenspaces are one dimensional.

(c) A has a left eigenvector and a right eigenvector with eigenvalue r, whose components are all positive.

(d) The only eigenvectors whose components are all positive are those associated with the eigenvalue r.

(e) The Perron–Frobenius eigenvalue is bounded above and below by the maximum and minimum row sums of A, respectively:

$$\min_i \sum_{j=1}^{m} A_{ij} \le r \le \max_i \sum_{j=1}^{m} A_{ij};$$

similarly r is bounded above and below by the maximum and minimum column sums of A respectively.

Exercise 1.2

Suppose an $n \times n$ matrix A is non-negative and the spectral radius of A given by the Perron–Frobenius eigenvalue is r. Let $\alpha \in (0, r)$. Then consider

$$(I - \alpha A)^{-1}.$$

Show that if this matrix is strictly positive then A is irreducible. Is the converse true? Show that if the matrix

$$\exp(A) = I + A + A^2/2! + A^3/3! + \cdots$$

is strictly positive then A is irreducible. Is the converse true? (In Section 2.2 we show how to define such functions of matrices without the need to use series, by exploiting the Jordan form of A.)

Exercise 1.3

Let A be an $n \times n$ non-negative matrix. Suppose A is written in block form:

$$A = \begin{pmatrix} 0 & B_1 \\ B_2 & 0 \end{pmatrix}$$

such that B_1 is $n_1 \times n_2$ ($n_1 + n_2 = n$) and B_2 is $n_2 \times n_1$, and both are strictly positive. Show that A is irreducible.

Such a block matrix could represent two segments of a population: say n_1 managers and n_2 workers, with the elements representing the workers' preferences for working for various managers, and managers' preferences for managing various workers, expressed on some strictly positive scales.

1.3 Singular Value Decomposition (SVD)

The singular value decomposition of a matrix is an extremely useful factorization of a general matrix that can be applied in various fields: matrix approximation, the definition of pseudo inverses, total least squares, and many other applications including signal processing and state space embedding methods.

Let M denote an $m \times n$ matrix whose entries are real or complex. The singular value decomposition of M is a factorization of the form

$$M = U\Sigma V^*$$

where U is an $m \times m$ unitary matrix; Σ is an $m \times n$ diagonal matrix with non-negative real numbers on the diagonal, and the $n \times n$ unitary matrix V^* denotes the conjugate transpose of the $n \times n$ unitary matrix V. The diagonal entries of Σ are known as the singular values of M. A common convention is to list the singular values in descending order.

The m columns of U and the n columns of V are called the left-singular vectors and right-singular vectors of M, respectively.

The ideas of singular value decomposition and the eigenvalue decomposition are closely related. In particular, we have both

$$MM^* = U(\Sigma\Sigma^*)U^* \quad M^*M = V(\Sigma^*\Sigma)V^*,$$

and these relations describe eigenvalue decompositions. Thus the left-singular vectors of M are eigenvectors of MM^*, and the right-singular vectors of M are eigenvectors of M^*M. Hence the non-zero singular values of M (found on the diagonal entries of Σ) are the square roots of the non-zero eigenvalues of both M^*M and MM^*.

The singular value decomposition can define the *pseudoinverse* of a matrix. Indeed, the pseudoinverse of the matrix M with singular value decomposition is

$$M^+ = V\Sigma^+U^*$$

where Σ^+ is the pseudoinverse of Σ, formed by replacing every non-zero diagonal entry by its reciprocal and transposing the resulting matrix. Working numerically one only 'inverts' those singular values greater than some small tolerance.

The singular value decomposition can be applied to any $m \times n$ matrix, whereas eigenvalue decomposition can only be applied to certain classes of square matrices.

In the special case that $m = n$ and M is a normal matrix, the spectral theorem says that it can be unitarily diagonalized using a basis of eigenvectors, so that it can be decomposed UDU^*, with a unitary matrix U and a diagonal matrix D. Thus if M is also positive semi-definite, then the eigenvalue decomposition is also a singular value decomposition. Yet the eigenvalue decomposition and the singular value decomposition will differ for all other matrices.

Exercise 1.4

Suppose A is normal and invertible. Then there is a unitary U such that $A = U \Lambda U^T$ and Λ is diagonal containing the eigenvalues of A. Let $f : \mathbb{R} \to \mathbb{R}$ be any function that is well defined at all of the eigenvalues of A. Define $f(A) = Uf(\Lambda)U^T$ where $f(\Lambda)$ is diagonal: with f applied to each corresponding element of Λ.

(a) Show that if Q is any polynomial $Q(x) = q_0 + q_1 x + \cdots + q_m x^m$ then
$$Q(A) = q_0 I + q_1 A + \cdots + q_m A^m.$$

(b) Similarly show that
$$Q(A^{-1}) = q_0 I + q_1 A^{-1} + \cdots + q_m (A^{-1})^m = UQ(\Lambda^{-1})U^T,$$
and

(c) that
$$Q(A)^{-1} = (q_0 I + q_1 A + \cdots + q_m A^m)^{-1} = UQ(\Lambda)^{-1}U^T.$$

Exercise 1.5

Suppose A is normal and its spectral radius is $\rho(A) < 1/\alpha$ for some $\alpha > 0$. Then consider $(I - \alpha A)^{-1} = U(I - \alpha \Lambda)^{-1}U^T$. Show that this is the geometric series:
$$S = I + \alpha A + \alpha^2 A^2 + \alpha^3 A^3 + \cdots$$

1.4 Graph Theory Preliminaries and Random Graphs

Now we turn to some basic definitions and concepts from graph theory. We shall use the terms 'network' and 'graph' interchangeably. There are many good books on modern graph theory [13].

A *graph* $G = G(V, E)$ is a set of vertices, V, usually of finite size, say n, and often labelled, say $V = \{v_1, \ldots, v_n\}$, and a set of edges, E, which is a given subset of $V \times V$ (the pairs of connected vertices).

If the graph is *directed* then the ordering of the pairs in E is important: if $(v_i, v_j) \in E$ then we say there is a directed edge from v_i to v_j. If the graph is *undirected*, then the ordering of the pairs in E is irrelevant. In that case each edge connects v_i to v_j, say, and vice versa. Graphs will be undirected in this chapter (unless we stipulate otherwise) with the extensions to undirected graphs being obvious for the most part.

The set of all possible undirected graphs on n vertices becomes very large as n becomes large. There are $n(n-1)/2$ possible edges, and each can be either present or absent. Hence there are $2^{n(n-1)/2}$ possible graphs.

A graph G will often be represented by its *adjacency matrix, A*. We assume that there is an enumeration of the vertices, so that $V = \{v_1, \ldots, v_n\}$; then A is a binary matrix such that the (ij)th element $A_{ij} = 1$ if $(v_i, v_j) \in E$, and $A_{ij} = 0$ otherwise. For an undirected graph the adjacency matrix is symmetric, so G is fully represented by the upper triangular part of A alone. In what follows we shall always rule out the possibility of edges connecting a vertex to itself (loops). So all of the main diagonal terms in the adjacency matrix will be zero.

Of course if we permute the enumeration of the vertices we have the same graph but with a different adjacency matrix. We are often interested in those properties of A that are invariant with respect to such permutation. These include spectral properties of A and matrices linearly related to A (such as the graph Laplacian), whether or not A is reducible, and also the canonical form for A.

Later we shall search for particular orderings or enumerations of the vertices so that A maximizes a given likelihood; that of being generated by an assumed model.

Let S denote the set of all possible $n \times n$ matrices that (i) are symmetric; (ii) have elements taking *real* values in [0,1]; and (iii) have a zero main diagonal. Then the space S has $n(n-1)/2$ degrees of freedom and is equivalent to the space $[0, 1]^{n(n-1)/2}$.

Let $\mathcal{A}$ denote the subset of S of all possible $n \times n$ adjacency matrices that (i) are symmetric; (ii) have elements that take *binary* values in $\{0, 1\}$; and (iii) have a zero main diagonal. Then the space $\mathcal{A}$ contains $2^{n(n-1)/2}$ elements, so as n increases it gets very large. If $n = 10$ there are more than thirty-five trillion graphs.

Sometimes in applications graphs are weighted, in which case A is replaced by a weighted version (containing real non-negative weights) for each edge and then it can be scaled so as to lie in S.

Here is the adjacency matrix for an undirected (unweighted) graph G:

$$A = \begin{pmatrix}
0 & 1 & 0 & 0 & 1 & 1 & 0 & 0 & 0 & 0 & 0 & 0 & 0 & 0 & 0 & 1 & 1 & 0 & 0 & 1 \\
1 & 0 & 1 & 1 & 1 & 1 & 1 & 0 & 0 & 0 & 1 & 1 & 0 & 0 & 0 & 0 & 0 & 1 & 1 & 0 \\
0 & 1 & 0 & 0 & 1 & 0 & 1 & 1 & 1 & 0 & 0 & 0 & 0 & 1 & 1 & 1 & 0 & 0 & 0 & 0 \\
0 & 1 & 0 & 0 & 0 & 1 & 1 & 1 & 0 & 1 & 0 & 1 & 0 & 0 & 0 & 1 & 0 & 0 & 0 & 0 \\
1 & 1 & 1 & 0 & 0 & 1 & 1 & 1 & 1 & 0 & 0 & 0 & 0 & 0 & 0 & 0 & 0 & 0 & 0 & 0 \\
1 & 1 & 0 & 1 & 1 & 0 & 1 & 0 & 1 & 0 & 0 & 0 & 1 & 1 & 1 & 0 & 0 & 1 & 1 & 0 \\
0 & 0 & 1 & 1 & 1 & 1 & 0 & 0 & 1 & 1 & 0 & 1 & 0 & 1 & 0 & 0 & 0 & 0 & 1 & 0 \\
0 & 0 & 1 & 1 & 1 & 0 & 0 & 0 & 1 & 1 & 1 & 0 & 1 & 0 & 0 & 0 & 0 & 0 & 1 & 0 \\
0 & 0 & 1 & 0 & 1 & 1 & 1 & 0 & 0 & 1 & 0 & 1 & 0 & 0 & 1 & 0 & 0 & 0 & 0 & 0 \\
0 & 1 & 0 & 1 & 0 & 0 & 1 & 1 & 1 & 0 & 0 & 0 & 1 & 1 & 0 & 1 & 1 & 0 & 0 & 0 \\
0 & 1 & 0 & 0 & 0 & 0 & 0 & 1 & 0 & 0 & 0 & 1 & 0 & 1 & 1 & 0 & 1 & 0 & 1 & 0 \\
0 & 0 & 0 & 1 & 0 & 0 & 1 & 1 & 1 & 0 & 1 & 0 & 1 & 0 & 0 & 0 & 0 & 0 & 1 & 0 \\
0 & 0 & 0 & 0 & 0 & 1 & 0 & 0 & 0 & 1 & 0 & 1 & 0 & 1 & 1 & 1 & 1 & 0 & 1 & 1 \\
0 & 0 & 1 & 0 & 0 & 1 & 1 & 1 & 0 & 1 & 1 & 0 & 1 & 0 & 0 & 1 & 1 & 1 & 1 & 0 \\
0 & 0 & 1 & 0 & 0 & 1 & 0 & 0 & 1 & 0 & 1 & 0 & 1 & 0 & 0 & 0 & 0 & 0 & 1 & 0 \\
1 & 0 & 1 & 1 & 0 & 0 & 0 & 0 & 0 & 1 & 0 & 0 & 1 & 1 & 0 & 0 & 0 & 1 & 1 & 1 \\
1 & 1 & 0 & 0 & 0 & 0 & 0 & 0 & 0 & 1 & 1 & 0 & 1 & 1 & 0 & 0 & 0 & 0 & 1 & 0 \\
0 & 1 & 0 & 0 & 0 & 1 & 0 & 0 & 0 & 0 & 0 & 0 & 0 & 1 & 0 & 1 & 0 & 0 & 0 & 0 \\
0 & 0 & 0 & 0 & 0 & 1 & 1 & 1 & 0 & 0 & 1 & 1 & 1 & 1 & 1 & 1 & 1 & 0 & 0 & 1 \\
1 & 0 & 0 & 0 & 0 & 0 & 0 & 0 & 0 & 0 & 0 & 0 & 1 & 0 & 0 & 1 & 0 & 0 & 1 & 0
\end{pmatrix}.$$

Many mathematical packages contain routines that allow us to visualize graphs. In Figure 1.1 we depict the graph represented by the adjacency matrix for an undirected graph G.

The *degree* of a vertex is the number of edges that connect to that particular vertex. The degree of vertex v_i is simply the ith row/column sum of the A. (Directed graphs would have an out-degree and an in-degree). The degree distribution, $P(d)$, represents the probability that a randomly chosen vertex has a particular degree, d. It is clear that the sum of the degrees of all of the vertices is equal to twice the number of edges.

The vertex degrees in the above example are given by (6, 9, 8, 7, 7, 11, 9, 8, 7, 9, 7, 7, 9, 11, 6, 9, 7, 4, 11, 4).

A *clique* is a graph where all of the edges are present: each vertex is connected to all others. The adjacency matrix for a clique contains all ones except for the main diagonal, which contains zeros (as we do not allow edges connecting a vertex to itself). We usually denote this adjacency matrix by *1*.

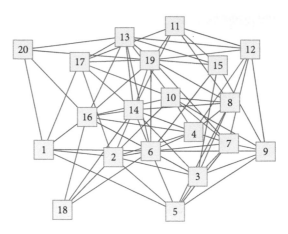

Fig 1.1 The graph G corresponding to adjacency matrix A.

If $G = (V, E)$, then the *complementary graph* is given by $G' = (V, E')$, where E' is the complement of E, and is the set of all admissible edges not in E. If A is the adjacency matrix for G, then $1 - A$ is the adjacency matrix for the complementary graph, G'.

When n gets large, graphs may contain information that is hidden. As an example, using the Chinese social network Weilink [80], Figure 1.2 shows the adjacency matrix for Weilink, where we have enumerated the individuals in the order they registered. What is the distribution of the size of any subgraphs that are cliques?

A random graph is simply an undirected graph where we have a probabilistic rule that decides whether each edge, or subsets of edges, is present or absent. We can contrast and manipulate random graphs by thinking about their representation by elements of $\mathcal{A}$.

An undirected *random graph* is defined by a probability distribution defined over the set of all possible graphs, $\mathcal{A}$. Random graphs can be thought of as *models* that are often used to generate observable instances of graphs with certain (expected or likely) properties. The probabilistic rule (the model) is a recipe that uses random processes to determine whether every edge is present (the elements in the upper triangular part of A). We may denote a random graph as a probability distribution $P(A|X)$, assigning a probability to each and every possible adjacency matrix. (As usual, X denotes 'everything else that we know *a priori*', before addressing A.) We must apply the model to determine the upper triangular part of A and then impose symmetry (while the main diagonal will always be zero).

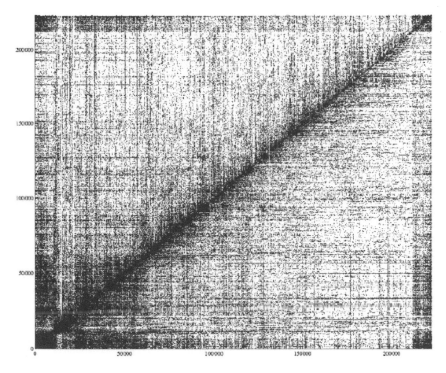

Fig 1.2 The adjacency matrix A for the first 225,000 members of Weilink [80].

For example, we might define a random graph by saying that each and every edge (in the upper triangular part of A) is present independently with the same given probability p. To generate such a graph's adjacency matrix we simply visit each and every edge in turn and include it with probability p (drawn independently for each edge), and then impose symmetry to complete A.

We may define the *expected value* of A as this first moment of the distribution: a real-valued matrix given by

$$\langle A|X\rangle = \sum_{A\in\mathcal{A}} AP(A|X).$$

Note the summation is over all possible values for A in $\mathcal{A}$. It is obvious that $\langle A|X\rangle$ is real and is in $\mathcal{S}$.

A random graph with distribution $P(A|X)$ is *edge independent* if, once all conditional information external to A is fixed and known, the additional knowledge of the existence or otherwise of any particular edge within A has no effect on than probability than any other edge in A.

This assertion is not as simple as it first looks. Consider the case where we are given $P(A|Q, X)$ and two edges are conditional on the same piece of conditioning information; the truth of some proposition, Q, that may be determined *a priori*, external to any details of A itself. Then, while the truth of Q is unknown, the knowledge of one edge being present can affect the probability that the other edge is present. In fact Bayes' Theorem says that the presence of one such edge would affect the probability that Q is true under these circumstances, and then such a change in probability for Q would affect the probability that other edges might be present. But 'edge independence' means that once the status of all Q's that are external to the graph are fully known (true or false), then under all scenarios the presence of any two edges in A may be determined by their conditional probabilities independently: we are asserting that the additional knowledge of the presence of one edge does not change the probability that the other is present. So if edges are independent yet conditional on the same piece of external conditioning information, Q, then they will be correlated over repeated tests (through this mutual dependence). Yet they are still independent once the truth of all prior external propositions, Q, is known.

Now if we know $\langle A|X \rangle$ and we also know or assume that it is edge independent then $\langle A|X \rangle$ has a ready interpretation. The (i, j)th element of $\langle A|X \rangle$, for $i < j$, is simply the independent probability that the (i, j)th edge is present in A. So we can generate an instance of A very easily by visiting each possible edge, $i < j$, and including it with the required probability, and then imposing symmetry.

By definition $P(A|X)$ is the probability that any A is generated by the corresponding random model. Since the edges are independent, for any A we can consider each edge, $i < j$, in turn. So we can express $P(A|X)$ as a product over every possible edge, $i < j$, of the probability that it is in A or is not in A, as required:

$$P(A|X) = \prod_{i=1, j=i+1}^{n} (\langle A|X \rangle)_{ij}^{A_{ij}} \, (1 - \langle A|X \rangle)_{ij}^{1-A_{ij}}.$$

Thus, if A is edge independent then $P(A|X)$ can be written directly in terms of $\langle A|X \rangle$.

If the edges of A are assumed to be non-independent then we will have to refer to the defining recipe (the defining model) for P in order to generate instances of A.

A Non-Independent Example

Consider the graph on $n = 20$ vertices where each closed triangle of vertices, (v_i, v_j, v_k), is included with an independent probability $p=0.02$. Thus the graph is a union of randomly generated triangles. Is this graph edge independent?

There are $m = n - 2$ possible triangles that contain each edge (v_i, v_j). Thus each edge has a probability of

$$1 - (1 - p)^m = 0.3049\ldots$$

of being present within this graph. Now suppose that we know that the edge (v_i, v_k) is present. What is the probability that edge (v_i, v_j) $(i \neq j)$ is present too? The triangle (v_i, v_k, v_j) is one of m triangles containing the edge (v_i, v_k). Thus (v_i, v_k, v_j) has an updated probability, q of being present as a result of this new information. In fact $q = p/(1 - (1 - p)^m) > p$. This last expression can be best calculated by applying Bayes' Theorem to update p because

$$\frac{q}{1-q} = \frac{P((v_i, v_k)|(v_i, v_j, v_k), X)}{P((v_i, v_k)|\text{not }(v_i, v_j, v_k), X)} \times \frac{p}{1-p} = \frac{1}{1 - (1-p)^{m-1}} \times \frac{p}{1-p}.$$

Now all of the other $(m - 1)$ triangles involving (v_i, v_j) still have a probability p of being present. So now (v_i, v_j) has a probability

$$1 - (1 - p)^{m-1}(1 - q) = 0.3372\ldots$$

of being present. Hence knowledge of one edge has changed our estimation of whether a second edge can be present.

Exercise 1.6

Suppose we form a network of people as follows: each time a new person joins a company they select exactly one of the existing employees at random (with equal probability) as a 'buddy'. Edges of the network represent the two-way buddy relationship. Nobody ever leaves the company and buddies are for a (corporate) lifetime. Is this an edge-independent network? What is the expected degree of the kth member of staff to join after $K \geq k + 1$ members of staff have joined? What happens as $K \to \infty$?

Exercise 1.7

In the army, each trouper reports to their own corporal, each corporal reports to their own sergeant, and so on. This type of graph is called a tree. Is it edge independent?

What about swans who mate for life?

What about men and women at a formal dance? Suppose there are ten separate dances on the card, where any man may dance with up to ten women, and any woman may dance with up to ten men. If no 'cutting in' is allowed, is it edge independent? What if the number of dances is very large? This is called a *bipartite* network.

A *walk* is a sequence of vertices such that the successive vertices are adjacent (connected by present edges). The length of the walk is the number of edges traversed (the length of the vertex sequence minus one). A closed walk is a *cycle*. A *path* is a walk between two vertices of a graph that visits each vertex at most once. A *Hamiltonian path* is a walk between two vertices of a graph that visits each and every vertex exactly once.

The following theorem gives one very important use of powers of the adjacency matrix of a graph.

Theorem 1.1. If A is the adjacency matrix of a graph G (with vertices $v_1, \ldots, v_n$), for an integer $r \geq 1$, then the (i,j)th-entry of A^r represents the number of distinct walks of length r from v_i to v_j in the graph.

Proof. This is simply counting. We prove this by induction.

The theorem is trivially true for $r = 1$. Suppose it is true for any r. Then for any i, let $n_k^{[r]} = (A^r)_{ik}$ denote the kth element of the ith row of A^r, which is the count of the number of walks from vertex i to vertex k of length r (by assumption). Now consider any vertex v_j; for any existing walk of length $r + 1$ from v_i to v_j there must be a penultimate vertex v_k such that v_k is adjacent to vertex v_j, and the previous part of the walk is itself a walk of length r connecting v_i to v_k. There are $n_k^{[r]}$ such walks (by assumption). Thus counting up all such walks we have (ranging over k):

$$n_j^{[r+1]} = \sum_{A_{kj} \neq 0} n_k^{[r]} = \sum_{k=1}^{n} n_k^{[r]} A_{kj} = \sum_{k=1}^{n} (A^r)_{ik} A_{kj} = (A^{r+1})_{ij}.$$

The result follows.

It is easy to see how this generalizes.

Theorem 1.2. Let A be an adjacency matrix of a graph G over vertices $v_1, \ldots, v_n$. Let B be the adjacency matrix of a graph G' over the same set of vertices. Then the (i, j)th-entry of $A^r . B^{r'}$ represents the number of distinct walks of length $r + r'$ from v_i to v_j, taking the first r edges from the graph G, and then r' edges from the graph G'.

A graph is *connected* if there is a walk between any two pairs of vertices.

The *distance* between any pair of vertices is the length of the shortest walk connecting them. The *median diameter* of a graph is the median distance between pairs of vertices. That is the distance between any pair of vertices is equally likely to be longer or shorter than the median diameter.

The Perron–Frobenius theorem introduced earlier applies to adjacency matrices that are irreducible. The condition for irreducibility is equivalent to connectedness (for all i and j there exist $k = k(i, j)$ such that $(A^k)_{ij} > 0$).

The *graph Laplacian*, denoted by Δ, is the symmetric matrix $D - A$, where $D = \text{diag}(d_1, d_2, \ldots, d_n)$ is the diagonal matrix of the vertex degrees.

We consider the graph Laplacian as well at the Laplacian for matrices in S in Section 1.5 and show that it is positive definite and contains other important information.

Exercise 1.8

What is the degree distribution for a clique, a circular lattice, and a lattice, all on vertices n?

Exercise 1.9

If some given $P(k)$ is the degree distribution for a graph G on vertices n, what is the degree distribution for the complementary graph G'?

Exercise 1.10

Suppose a graph G has a degree distribution given by $P(k)$. Suppose that we are told a vertex v is connected to another vertex, say v_0. What is probability, $\tilde{P}(k)$, that v has degree k, given this extra information?

Erdős–Rényi Random Graphs

In graph theory, the Erdős–Rényi random graph model, named for Paul Erdős and Alfred Rényi, denoted by $G_{(n,p)}$, is constructed over n vertices where each possible edge is included in the graph with an independent probability equal to p. The expected number of edges in $G_{(n,p)}$ is thus $pn(n-1)/2$. The expected value of the adjacency matrix is given by

$$\langle A|X \rangle = p\mathbf{1} \in S.$$

The distribution for the degree, m, of any particular vertex is given by a binomial distribution:

$$P(\text{vetex degree} = m|X) = C_m^{n-1} p^m (1-p)^{n-1-m}.$$

In this model when n is large and p is small, so the expected degree $\lambda = (n-1)p$ is much less than n, the binomial distribution can be approximated by a Poisson distribution:

$$P(k|\lambda, X) = \frac{\lambda^k e^{-\lambda}}{k!}.$$

Exercise 1.11

Show that both the mean and variance of $P(k|\lambda, X) = \frac{\lambda^k e^{-\lambda}}{k!}$ are equal to λ.

Random Lattices

A lattice is a graph where the vertices are enumerated (v_i, $i = 1, 2, \ldots, n$) and all k near-neighbours are connected by edges. Thus $(v_i, v_j) \in E$ if and only if $0 < |i - j| \leq k$. The adjacency matrix for such a lattice on n vertices contains ones on the first k sub-diagonals as well as first k super-diagonals. Let us denote this matrix (and its graph) by $L_{(n,k)}$.

Sometimes we bend the enumeration of the vertices around into a circle, and produce a circular lattice, with $(v_i, v_j) \in E$ if, and only if, $0 < |(i-j)| \mod (n)k$. The adjacency matrix for such a circular lattice on n vertices contains ones on the first k and last k sub-diagonals, as well as the first k and last k super-diagonals. Let us denote this matrix $CL_{(n,k)}$.

Random lattices are constructed by starting with any lattice (linear or circular) and retaining each edge independently with a certain probability p. They have an expected adjacency matrix that can be written $\langle A|X \rangle = pA'$ whether A' is the adjacency matrix for a lattice. Intuitively this means that every edge that is in the full lattice represented by A' is included in the random lattice with independent probability p. Edges not in the lattice A' are thus not in the corresponding random lattice.

Exercise 1.12

1. Name the graphs in Figure 1.3.

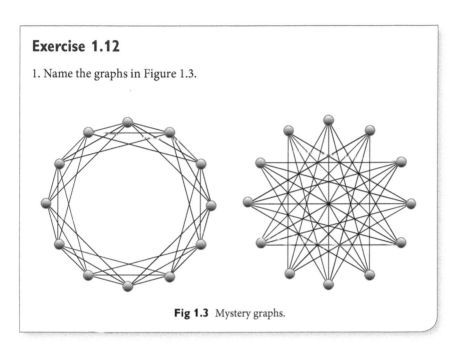

Fig 1.3 Mystery graphs.

Range-Dependent Random Graphs

Erdős–Rényi graphs are independent of the particle enumeration of the verities. Let us consider a class of undirected random graphs that have a model that depends critically on the chosen enumeration (so vertices are not indistinguishable) [52]. Later we shall try to determine the enumeration (ordering) from given observations as an inverse problem.

For a given enumeration of the vertices, the *range* of the possible edge connecting vertex i and vertex j is simply $|j - i|$. Assuming each edge is independent of others, we shall generate such an adjacency matrix by visiting each edge ($j > i$) and including it or not with the probability that is conditionally dependent on the range. Specifically we will have a function $f : Z^+ \rightarrow [0, 1]$; usually it will be decreasing so that longer range edges are successively less likely to be observed than shorter range edges. Then we impose symmetry and a zero main diagonal to complete the adjacency matrix, $A \in \mathcal{A}$.

The expected value, $\langle A|X \rangle$, of our adjacency matrix is a Toeplitz matrix in S, with all of the elements on the kth super-diagonal (and by symmetry on the kth sub-diagonal) equal to $f(k)$. The random adjacency matrix is equivalent to a probability distribution defined over all possible adjacency matrices, $\mathcal{A}$. Formally $\mathcal{A}$ is the set of all symmetric matrices containing binary values, with a zero main diagonal: there are $2^{n(n-1)/2}$ elements in that set. Moreover, since the edges of our random undirected adjacency matrix are independent (by assumption), then for any candidate $A \in \mathcal{A}$, we may write down the probability distribution:

$$P(A|X) = \prod_{i=1, j=i+1}^{n-1, n} f(j-i)^{A_{ij}} (1 - f(j-i))^{(1-A_{ij})}.$$

This form is particularly elegant since there are exactly $n(n-2)/2$ terms in the product with each A_{ij} picking out the desired probability from the corresponding pair. Similar forms are used to derive maximum likelihood equations for logistic regression methods. This is a very useful construct indeed when dealing with binary variables (such as the A_{ij} here).

Range-dependent random graphs can also possess the small world property introduced in the small world Random Graphs section [52].

If we allow $n \rightarrow \infty$, the range-dependent graphs can be considered via a generating function.

The *generating function* for a graph is parameterized by its degree distribution, $P(k|X)$ (for degree k). It is a function of a real variable $x \in [0, 1]$ and is defined by

$$\mathcal{G}(x) = \sum_{k=0}^{\infty} P(k|X)x^k,$$

where $P(k|X)$ denotes the probability that a (uniformly) randomly chosen vertex has degree k.

$\mathcal{G}(x)$ is very useful to theorists since the moments of the degree distribution relate directly to the values of $\mathcal{G}$ and its derivatives at $x = 1$, while the derivatives of $\mathcal{G}$ at zero are trivially related to the values in the degree distribution.

Consider the range dependent graph above, where $n \rightarrow \infty$. Fix on a random vertex, say v_0. Then edges connect v_0 to other possible vertices v_k, $k = \ldots, -2, 1, 1, 2, \ldots$, with corresponding probability $f(|k|)$.

Consider the product

$$\prod_{j=-\infty,\,j\neq 0}^{\infty} (xf(|j|) + (1 - f(|j|))).$$

The x^k term represents the sum over all of the distinct ways that v_0 can have exactly k adjacencies. Each of these has exactly k edges present (giving a single $f(|j|)x$ factor) with the corresponding probability $f(|j|)$. Hence this product is equal to $\mathcal{G}(x)$. Simplifying we can write

$$\mathcal{G}(x) = \prod_{j=1}^{\infty}(xf(j) + (1 - f(j)))^2.$$

Then $\mathcal{G}(1) = 1$ and the derivative $\mathcal{G}_x(1) = 2\sum_{j=1}^{\infty} f(j)$ is the expected degree.

Exercise 1.13

Show that $\mathcal{G}_x(1) = 2\sum_{j=1}^{\infty} f(j)$.

Exercise 1.14

If $f(k) = b/(\pi k)^2$, then by definition

$$\mathcal{G}(x) = \prod_{j=1}^{\infty} \left(1 - \frac{b}{(j\pi)^2}(1 - x)\right)^2.$$

In fact it is known (but not well known) that this infinite product can be rewritten. We have

$$\mathcal{G}(x) = \frac{\sin^2 \sqrt{b(1 - x)}}{b(1 - x)}.$$

Find the mean degree $\mathcal{G}'(1)$ and the first four P_k's.

Small World Random Graphs

Lattices and random lattices have a high degree of localized clustering. For example, if you know Suzie, and Suzie knows Ramona, then there is a high probability that you also know Ramona. In their work on small world graphs, Watts and Strogatz [124, 125] used this idea to define the Watts–Strogatz clustering coefficient which measures how transitive adjacency is. It is intuitive for social networks: on Facebook you can see how many friends you have in common with any friend of your own.

The Watts–Strogatz clustering coefficient, C, measures the fraction of the friends of friends that are also friends. We count the number of ordered vertex triples (v_i, v_j, v_k) such that both v_i and v_k are adjacent to v_j; and then C measures the fraction of these for which v_i and v_k are also adjacent.

For example, for $G_{(n,p)}$ each edge is independent and hence the probability that any pair v_i and v_k are adjacent is simply p. So we must have $C = p$ for such graphs.

Consider a large lattice, $L_{(n,k)}$ for $n >> k$. Ignoring edge effects (occurring only at the first and last k vertices), each edge has degree $2k$. Thus each vertex has $2k(2k - 1)/2$ distinct pairs of connected vertices. Of those pairs how many are themselves directly connected? $3(k - 1)k/2$. Thus for $L_{(n,k)}$ and n large, we have $C = (3k - 1)/(4k - 2)$. For large k this approaches 3/4. For a clique we would have $C = 1$, so $C = 3/4$ is relatively high. (see Exercise 1.15.)

The attraction of lattices in having a relatively high clustering coefficient is obvious when we think about social networks. However, they lack an important property exhibited within social networks which more random networks possess.

How far away is one vertex from another? What is the expected distance between any pair of vertices? For the lattice $L_{(n,k)}$ it is approximately their distance of separation along the lattice divided by k. This will typically be large when n is large.

But early ground-breaking work with social communication networks showed that six degrees of separation were often enough, as demonstrated in experiments in the sixties by Milgram [104] (though in these days of Facebook only four to five degrees are sufficient [7]). So lattices are not such good models of social networks after all.

Watts and Strogatz noted this, and suggested a simple way to retain the properties of a lattice, while including some longer range edges that would drastically reduce the distances between pairs of vertices and hence the median diameter of the graph as a whole. Effectively their construction starts out from a lattice, say $CL_{(n,k)}$, and then it partly rewires the lattice by deleting some of its edges and replacing these with random edges that were not in the lattice previously.

This process produces a graph, G, with adjacency matrix, A, that is conditionally dependent on that of the initial lattice. In our notation above, we considered writing a new graph conditional on some given graph. Hence we will take an algebraic approach to analyse the Watts–Strogatz construction. Normally it is introduced simply via the rewiring algorithm.

We generate a small world graph as a random graph with independent edges, so it is enough to merely specify its expected adjacency matrix A:

$$\langle A | CL_{(n,k)} \rangle = pCL_{(n,k)} + q(1 - CL_{(n,k)}).$$

Here p and q are constants in $[0,1]$. We shall choose them so that the expected number of edges within A (and thus the average vertex degree) remains the same as that in $CL_{(n,k)}$: there are nk edges in total (since all vertices have degree $2k$, and sum of the degrees is equal to twice the number of edges). So we relate p and q via

$$nk = pnk + q(n(n - 1)/2 - nk).$$

Then the expected number of edges in $\langle A | CL_{(n,k)} \rangle$ is exactly nk. Note that if $q = 0$, then $p = 1$, and $\langle A | CL_{(n,k)} \rangle = CL_{(n,k)}$; while if $p = q$ we must have $q = q^* = 2k/(n - 1)$ and then $\langle A | CL_{(n,k)} \rangle = q^* \mathbf{1}$ and A is an Erdős–Rényi random graph, $G_{(n,q^*)}$.

We can think of q as a parameter varying continuously between zero and q^*. As it varies, the resulting graph goes through two phase changes. The first follows from the fact that the clustering coefficient, C, must fall from a relatively high value for the lattice (when $q = 0$, perhaps close to 3/4) down to $q^* = 2k/(n - 1) \ll 1$. The second follows from the fact that the median diameter starts high for the lattice, at approximately $n/4k$, but it falls to that of the Erdős–Rényi random graph. In Figure 1.4 we show such an example where q increases.

Watts and Strogatz showed that these two phase changes do not happen together. In particular, the significant fall in the median diameter happens for lower values of q than does the significant fall in the clustering coefficient. Between the two transitions lies the small world zone suitable for modelling social networks. The median diameter is small (so that something like six degrees of separation could be sufficient) while the clustering coefficient remains high, and adjacency is highly transitive. Graphs in that zone are small world graphs.

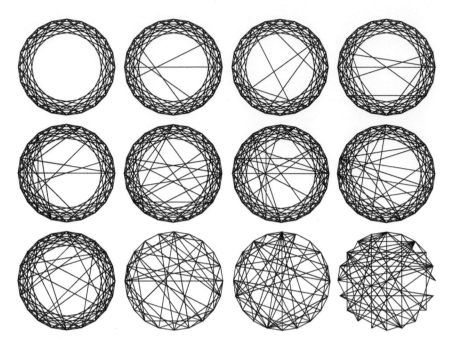

Fig 1.4 $\langle A|CL_{(n,k)}\rangle = pCL_{(n,k)} + q(1 - CL_{(n,k)})$ increasing q, where $p = 1 + q - q(n-1)/2k$ and $(n,k) = (20,4)$.

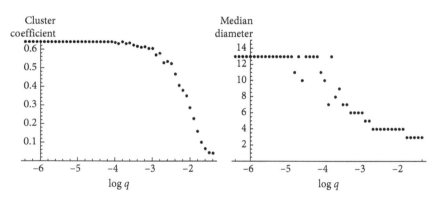

Fig 1.5 Cluster coefficient (left) and median diameter (right) against $\log q$.

For example, if we set $n = 200$, and $k = 4$, we have 800 edges in $CL_{(200,4)}$ (out of 19,900 possible edges on 200 vertices). Then $p = q$ when $q = q^* = 8/199$. In Figure 1.5 we show how the clustering coefficient and the median diameter vary as q varies. The small world zone is approximately where $q \in [10^{-3.5}, 10^{-2.5}]$, and C is still large while the diameter has fallen.

Exercise 1.15

Consider a large lattice, $L_{(n,k)}$, for $n >> k$, as above. Let us calculate the Watts–Strogatz clustering coefficient for this graph. Each vertex has $2k(2k-1)/2$ distinct pairs of connected vertices. Of those pairs, show that $(3k-1)k/2$ are directly connected forming a triangle.

Hint: imagine the adjacency matrix for the subgraph containing just the $(2k+1)$ vertices centred some chosen v_i (the vertices $v_{i-k}, \ldots, v_{k+i}$), and then count the number of zeros within the upper triangle and take this away from the size of that upper triangle less the $2k$ terms that appear within the middle row and column corresponding to vertex v_i s adjacencies.
What is C?

Exercise 1.16

Let A be the adjacency matrix for any graph. Let N_2 denote the number ordered vertex triples (v_i, v_j, v_k) such that both v_i and v_k are adjacent to v_j. Show that N_2 is the sum of the off-diagonal terms in A^2. Let N_3 denote the number ordered vertex triples (v_i, v_j, v_k) such that both v_i and v_k are adjacent to v_j, and v_i and v_k are also adjacent. Deduce that N_3 is the sum of the diagonal terms in A^3. Hence we have the Watts–Strogatz clustering coefficient: $C = N_3/N_2$. This provides a useful way to calculate C.

Scale-Free Networks and Preferential Attachment

Scale-free networks have a degree distribution that is of the form $P(d|X) \sim d^{-\gamma}$. Here the positive exponent γ is usually taken to be between 2 and 3.

Notice that if γ is less than 2 then the mean degree is ill-defined. For γ in (2,3] $d^2 P(d|X) \sim d^{2-\gamma}$ is not summable, so, though there is a finite mean (the expected degree), the distribution does not have a finite variance. It has a fat tail. Such distributions appear within many modern applications of stochastic processes where the law of large numbers fails (that is, beyond diffusion processes, such as within stochastic models of stock price movements and Levy flights).

A distribution of the form $P(d|X) \sim d^{-\gamma}$ is often referred to as a Pareto distribution or power law. Recent interest in scale-free networks all started in 1999 with seminal work by Barabasi and Albert who mapped the connectivity of a portion of the internet [11], and coined the term 'scale-free network' to describe the class of networks. Importantly, they also proposed an intuitive model to explain the appearance of such a power-law distribution, which is called *preferential attachment*. Analytic solutions for this mechanism were rigorously obtained [14]. This particular mechanism only produces a specific subclass of the networks within the scale-free class, and alternative mechanisms have been discovered since.

Consider the situation where a graph grows by having a single vertex added to the existing graph over each unit of time that is attached to exactly m of the existing vertices. The probability that it becomes attached to any vertex is assumed to be given by the relative degree of the vertex: that is, the vertex degree divided by the sum of all of the vertex degrees. Note that the sum of all vertex degrees is simply twice the number of edges.

Suppose at $t = 0$ we start with a very small number, say n_0, of unconnected vertices, and that new vertices are added randomly at an expected rate of one per unit time. After time t we will have added approximately t vertices and there will be approximately mt edges connecting the $t + n_0$ vertices.

Let $d_i(t)$ denote the expected degree of the ith vertex, the one that was added at around time $t = i$.

Then for $t \geq i$, $d_i(t)$ grows as successive vertices (and hence edges) are added, with vertex i winning new attachments at a rate proportional to its current relative degree. We have

$$\frac{d}{dt}d_i(t) = (\text{rate at which new edges are added}) \times P(\text{new edge attaches to vertex } i),$$

so that

$$\frac{d}{dt}d_i(t) = m\frac{d_i(t)}{2mt} = \frac{d_i(t)}{2t},$$

(since at time t the sum of all vertex degrees is twice the number of edges), and it must also satisfy the initial condition $d_i(i) = m$. So directly we have

$$d_i(t) = m(t/i)^{1/2}.$$

Hence for any value d chosen larger than m at time t, we will have $d_i(t) < d$ if, and only if, $i > tm^2/d^2$. So there are approximately $n_0 + t - tm^2/d^2$ vertices with degrees less than d. Therefore there is a fraction

$$1 - \frac{tm^2}{d^2(t + n_0)}$$

of the vertices that have degrees less than d. For large time, the distribution forgets n_0 and this is simply

$$1 - m^2/d^2.$$

Note that this does not make sense for $d < m$, since all vertices have degrees greater than or equal to m.

This is the cumulative distribution for the vertex degrees (valid for $d \geq m$). The density distribution (the derivative of the cumulative with respect to d) is thus $P(d|X) \sim d^{-3}$.

One often encounters statements about whether particular data sets result in scale-free behaviour or not, based on observed degree distributions and so on. But the usefulness of this is really rather questionable. In applications you have the data that you observe: you cannot change it. The most helpful element of this type of classification may be to contrast distinct data sets, and seeing some way that they are not structurally the same—even when others claim that they are. Yet once you have a given network, you must move on to other questions such as those addressed in this chapter and in other chapters (for social networks).

1.5 Similarity Matrices, Random Matrices, and Clusters

Suppose that we have n observations of some events, $\mathbf{x}_i$, that are vectors of real, integer, and categorical attributes. Let us assume that $\mathbf{x}$-space (the set of all possible values) has a metric, denoted by μ. Then one could define an $n \times n$ similarity matrix A with ijth element given by

$$A_{ij} = \mu(\mathbf{x}_i, \mathbf{x}_j).$$

Then A is symmetric and non-negative with a zero main diagonal. In some applications μ might be binary valued, in which case A is the adjacency matrix for the corresponding undirected network.

We would like a strategy to perform a *soft block-diagonalization* for A by permuting its rows and columns. That is, a method to bring together indices corresponding to subsets of mutually similar observations that should place the larger pairwise similarities close together within the diagonal blocks; and the smaller pairwise similarities away from the main diagonal, outside of those blocks, representing less 'local' block-to-block associations.

In all of the following applications, we shall assume that μ takes values only in $[0, 1]$. This is easily achieved by normalizing A.

This assumption will be very useful for us, as it allows us to interpret each element A_{ij} as being drawn from a real random process, defined over $[0,1]$, whose defining parameters depend on (i, j). Of course, by definition, A is symmetric, so we only need to model A_{ij} for index pairs above the diagonal $(j > i)$ and then we impose symmetry. In this way we shall think of the matrix A, or at least permutations of A, as realizations of $n(n - 1)/2$ suitable random processes—one for each element.

Again let S denote the set of all possible $n \times n$ *similarity matrices*, that (i) are symmetric; (ii) take values in $[0,1]$; and (iii) have a zero main diagonal. Then the space S has $n(n - 1)/2$ degrees of freedom and is equivalent to the space $[0, 1]^{n(n-1)/2}$. Clearly A and any permutation of A (permuting the rows and columns in the same way) lie in S. As before we denote by $\mathcal{A}$ the subset of S containing only binary values. Elements of $\mathcal{A}$ are thus *adjacency matrices* for undirected networks on n enumerated vertices.

A *random matrix* is a probability density distribution defined over a suitable set of matrices.

A *random symmetric similarity matrix* is thus equivalent to a probability density distribution defined over S, given by $P(R|X) \geq 0$ for all $R \in S$, where $\int_{R \in S} P(R|X)\, dR = 1$. The form of P may well be conditional on other parameters (that are hidden here) which describe the details of the distribution. Here, as usual, X stands for 'everything else that we as modellers know *a priori*'. In practice the random matrix only needs to determine the upper triangular part of $R \in S$, since the rest is completed by enforcing symmetry.

Suppose we are given some similarity matrix A. We might assume that some, yet to be determined, permutation of A is drawn from a suitable random symmetric similarity matrix (defined over S). Then we shall try to find that permutation, along with any other parameters used in the definition of the particular random matrix, through a maximum likelihood approach.

There is a huge research interest in random matrices which has grown over the past few years, in particular concerning properties of their spectra. In the analysis here we shall exploit random matrices so as to define suitable classes of matrices from which some permutation of A is assumed to be drawn. This will in turn allow us to find a suitable candidate permutation and thus perform the desired soft diagonalization of A into the permuted form.

Of course attributes of the spectrum of A are independent of any permutation, and it turns out the information that we will need is contained within some specific eigenvectors.

The treatment here applies both to well-defined similarity matrices in $\mathcal{S}$ and to the adjacency matrices in $\mathcal{A}$. In the latter case the terms 'similar' and 'similarity' should be interpreted as referring to the degree of mutual association, as represented by the networks (and where relevant its weights).

The Laplacian

Suppose that A is in $\mathcal{S}$ and let $\mathbf{s} = (1, 1, \ldots, 1)^T \in \mathbb{R}^n$. Then the vector $A\mathbf{s} = (d_1, d_2, \ldots, d_n)^T$ contains the row (and hence the column) sums of A. Let $D = \text{diag}(d_1, d_2, \ldots, d_n)$ be the diagonal matrix of A's row sums.

The *Laplacian* associated with A, denoted by Δ, is the symmetric matrix

$$\Delta = D - A.$$

Note the sign convention here: it is inherited from that of the discrete graph Laplacian (continuum mechanicists might wish to write Δ as the negative of this).

Now, for any $\mathbf{w} = (w_1, \ldots, w_n)^T \in \mathbb{R}^n$,

$$\mathbf{w}^T(D-A)\mathbf{w} = \mathbf{w}^T D\mathbf{w} - \mathbf{w}^T A\mathbf{w} = \sum_{i=1}^{n} d_i w_i^2 - \sum_{i,j=1}^{n} A_{ij} w_i w_j = \sum_{i=1}^{n} \sum_{i=j}^{n} A_{ij} w_i^2 - \sum_{i,j=1}^{n} A_{ij} w_i w_j.$$

But using the symmetry of A, we have

$$\sum_{i,j=1}^{n} A_{ij} w_i^2 = (1/2) \sum_{i,j=1}^{n} (A_{ij} w_i^2 + A_{ji} w_j^2) = (1/2) \left(\sum_{i,j=1}^{n} A_{ij}(w_i^2 + w_j^2) \right).$$

Hence

$$\mathbf{w}^T(D-A)\mathbf{w} = (1/2) \left(\sum_{i,j=1}^{n} A_{ij}(w_i^2 + w_j^2) - 2 \sum_{i,j=1}^{n} A_{ij} w_i w_j \right) = (1/2) \sum_{i,j=1}^{n} (w_i - w_j)^2 (A)_{ij}.$$

Thus

$$\mathbf{w}^T(D-A)\mathbf{w} = 1/2 \sum_{i,j=1}^{n} (w_i - w_j)^2 A_{ij}.$$

Thus Δ is always positive-semidefinite, and if $\mathbf{w}$ is an eigenvector of Δ corresponding to the zero eigenvalue, then its components must be the same for any pair (i,j) where $A_{ij} > 0$, and thus any pair for which $A_{ij}^k > 0$ for some power k.

Let the eigenvalues of $\Delta = D - A$ be denoted by $\lambda_0 \leq \lambda_1 \leq \ldots \leq \lambda_{n-1}$. Then we have deduced that (a) λ_0 is always zero because every Laplacian matrix has an $\mathbf{s}$ eigenvector and (b) the multiplicity of zero as an eigenvalue is 1 if A is irreducible.

The smallest non-zero eigenvalue of Δ is called the *spectral gap* or the *Fiedler eigenvalue*.

Note that this theory applies equally to similarity matrices in $\mathcal{S}$ and to adjacency matrices in $\mathcal{A}$. In fact the multiplicity of zero as an eigenvalue of the graph Laplacian is exactly equal to the number of connected components of a graph, since this theorem ensures that there is a corresponding eigenvector that is constant (and equal to one say) on each isolated sub-component, and zero elsewhere. If zero is a simple eigenvalue, the graph is connected and A is irreducible.

Range–Dependent Random Matrices

Next we introduce a class of random matrices that will be useful in providing a framework for optimization, and thus inverse problems.

A real Wigner matrix is a symmetric matrix where all of the elements are on or above the main diagonal and are independent random processes. The usual definition of Wigner matrices demands that all elements are identically distributed, but this is not quite what we need here. Rather we shall consider a class of random similarity matrices where (a) the main diagonal is always zero and (b) the elements along each super-diagonal are independently and identically distributed.

Condition (a) means that such random matrices can be thought of a probability distributions over the set $\mathcal{S}$, introduced above.

The *range* between any pair of indices i and j is simply the distance $|i - j|$. So the range is equal to k for those elements on the kth super-diagonal. Condition (b) simply says that elements on the kth super-diagonal are independently and identically distributed according to some random process that may depend on the range k.

We shall call these real symmetric *range-dependent random matrices*.

Let $g(z|\theta)$ denote some biased probability density function that is defined over real variable $z \in [0, 1]$, conditional on some biasing parameters, θ. Suppose

$g(z|\theta)$ has an expected value, denoted by $f(\theta)$, that is necessarily in [0,1]. Then for θ (and thus $f(\theta)$) fixed, we have

$$g(z|\theta) \geq 0, \quad z \in [0,1], \quad \int_0^1 g(z|\theta)\,dx = f(\theta).$$

For example, we could have $g(z|\theta) \propto z^a(1-z)^b$ where $a > 0$ and $b > 0$ are constant, satisfying $f = (a+1)/(b+2)$ (and $\theta = (a,b)^T$), so that g yields real numbers distributed in [0,1] with mean value f. Or else we might have $g(z;f) = (1-f)\delta(z) + f\delta(z-1)$ and $\theta = (f)$: here $\delta(z)$ denotes the Dirac, point mass, distribution at zero, so that g yields binary values with mean value f.

To define a range-dependent random matrix we simple allow $\theta = \theta(k)$ where $k = j - i$ is the range of the (ij)th element $(j > i)$.

Now given the definition for g and a parameter function, $\theta(k)$, defined for all integer ranges $k = 1, \ldots, n - 1$, we sample a matrix $R \in S$ by drawing each upper triangular element of R independently at random from our biased probability density distribution, g:

$$R_{ij} \sim g(z|\theta(j - i)), \quad j > i, \qquad (1.1)$$

and then imposing that R be symmetric with a zero main diagonal.

We shall also assume that $\theta(k)$ is defined in such a way that the corresponding expected value for $g(z|\theta(k))$, given by $f(k)$, is a monotonically decreasing function of k in [0,1].

The *element independence* assumption, that each upper triangular element is drawn independently from the appropriate distribution, is important here as it will allow us to multiply probabilities over the independent elements. In fact, given any $R \in S$ we can write

$$P(R|X) = \prod_{i=1, \, j=i+1}^{n-1, \, n} g(R_{ij}|\theta(j - i)).$$

Thus the element independence assumption, and the assumed full knowledge of $g(z|\theta(k))$, completely determines the corresponding probability distribution, P over S.

Note that the separate elements on the kth super-diagonal are all drawn from the same distribution, $g(z, \theta(k))$, and all have an expected value given by $f(k)$. If n is large, the central limit theorem tells us that the mean of the elements in R along the kth super-diagonal is distributed according to a Gaussian distribution about its expected value $f(\theta(k))$.

The expected value of our range-dependent random matrix is a real Toeplitz matrix, with zero main diagonal and equal to $f(\theta(k))$ along the k super- and sub-diagonals.

A particularly simple form for g is the following negative *exponential distribution*:

$$g(z|\theta(k)) = (\eta k^2)e^{-z\eta k^2}. \tag{1.2}$$

Here $\theta = (\eta k^2)$, where η is a constant in $[0,1]$. In this case, $f(k) = 1/(k^2\eta)$ (which is also the standard deviation).

Henceforth we shall employ g given in (1.2) in order to define our class of random similarity matrices over S, via 1.1, as these (decaying) random range-dependent matrices.

Now we return to our clustering (permutation) problem.

1.6 Inverse Problems via Range-Dependent Random Matrices

The class of range-dependent random matrices above lead us very naturally to consider some *inverse problems*.

Whenever a scientist seeks to impose some structure on a collection of observed data, he or she learns something. A wonderful example is provided by the periodic table of elements: by wrapping elements around with respect to their atomic numbers with a period of eight, one finds that elements with similar properties form up the columns, instantly validating the idea. This told us something about the structure of atoms, yet it also enabled the prediction of properties of the elements that had still to be observed.

So it may be with similarity matrices and adjacency matrices. By forcing a hypothetical model structure onto them and reordering the elements so they might be generated according to that model's principles, we may infer transitive properties between co-located elements, and especially between known and unknown elements.

Given data from a (large) symmetric similarity matrix A in S, how may we best represent that data as being generated by one of our class of range-dependent random matrices? We shall do so by finding a permutation of A which makes the observed matrix most likely to have been generated, since the given indices may be allocated arbitrarily, or even have been maliciously shuffled to hide this fact. In doing so, the expected value $f(k) = \langle g(.|\theta(k))\rangle$

is decreasing with k; we shall naturally permute A so that its larger values are closer to the main diagonal, where the stochastic processes are most likely to have generated them. Thus our permutation will achieve a soft block-diagonalization.

Let $\mathbf{q} = (q_1, \ldots, q_n)^T$ denote some permutation function for the index set $i = 1, \ldots, n$. Then since we assume independence of the individual superdiagonal matrix elements within our random matrix model, the probabilite of observing the data, given the permutation $\mathbf{q}$, can be formed by the product of the probabilities for each element, after the indices have been permuted. We have the conditional 'likelihood':

$$\mathcal{L} \equiv P(A|\mathbf{q}, X) = \prod_{i<j} g(A_{ij}|\theta(|q_i - q_j|)).$$

Thus if we seek $\mathbf{q}$ (and any other parameters used within the definition of g), then we should simply maximize the likelihood $\mathcal{L}$. By Bayes' Theorem (see the Appendix) this is equivalent to maximizing (finding the modal values) of the posterior conditional distribution for the unknowns, given the observations, and assuming a uniform prior over the set of permutations.

Substituting from (1.2) we have

$$\mathcal{L} = \prod_{i<j} \eta(q_i - q_j)^2 e^{-A_{ij}\eta(q_i - q_j)^2}.$$

Taking logs we have

$$\log \mathcal{L} = \left(\sum_{i<j} \log(\eta(q_i - q_j)^2) \right) - \eta \sum_{i<j} A_{ij}(q_i - q_j)^2.$$

But the first term is a constant and is independent of $\mathbf{q}$. So if we seek $\mathbf{q}$ so as to maximize $\log \mathcal{L}$ then we should minimize $\log \tilde{\mathcal{L}}$ given by

$$\log \tilde{\mathcal{L}} = \sum_{i<j} A_{ij}(q_i - q_j)^2 = \frac{1}{2} \sum_{i,j=1}^{n} A_{ij}(q_i - q_j)^2.$$

Now, by the earlier result, if $\Delta = D - A$ denotes the Laplacian for A, then we have

$$\log \tilde{\mathcal{L}} = \mathbf{q}^T \Delta \mathbf{q}. \tag{1.3}$$

If n is large, we might search for the minimum over the $n!$ permutations, though it may take a very long time. So instead we shall pursue an alternative course of action.

Of course (1.3) relies on the particular form that we have chosen in (1.2), with its k^2 range dependence. If we wish to choose other definitions for g (for modelling reasons), then it is often possible to find parameters in (1.2) that produce a nearby distribution, especially if we need sparse matrices, when only the small-k terms are important. The advantage of our choice will be clear as we proceed using algebra as our chosen weapon to optimize $\tilde{\mathcal{L}}$ rather than face a large combinatorial search.

Next we make a huge simplifying step. We *relax* the problem to consider a nearby problem that has a much easier solution. Above we sought a vector $\mathbf{q}$ taking integer values (and indeed all integer values in $\{1, \ldots n\}$) and then we wished to reorder the vertices in the increasing order given by the corresponding elements of $\mathbf{q}$ (the reordering equivalent to making the permutation $\mathbf{q}$).

Notice that if we add any fixed constant onto all elements of $\mathbf{q}$, or we multiply $\mathbf{q}$ by any positive constant, then this has no effect on the reordering. We just reorder by sorting vertices by the values of the corresponding elements, q_i.

So let us solve a relaxed version of this last minimization problem by allowing $\mathbf{q}$ to contain real values. This will give a sub-optimal solution, but hopefully not be so far from the actual solution. By relaxing the problem we avoid the huge combinatorial search and we can get to the solution to the new problem in one algebraic step. So we trade rigorous optimality for a very low cost algorithm.

We will minimize (1.3) with respect to $\mathbf{q} \in \mathbb{R}^n$, and we will reorder the vertices in increasing order of their corresponding real elements of $\mathbf{q}$. Now, since multiplying the solution $\mathbf{q}$ by any positive constant, or adding a fixed constant onto all elements, makes no difference, we must find a real $\mathbf{q}$ to satisfy

$$\min_{\|\mathbf{q}\|=1 \, \mathbf{q}.\mathbf{s}=0} \mathbf{q}^T \Delta \mathbf{q}.$$

Here $\mathbf{s} = (1, 1, \ldots, 1)^T$, which is the eigenvector of Δ corresponding to an eigenvalue of zero.

This minimum is achieved at the Fiedler eigenvalue (the second smallest eigenvalue), and $\mathbf{q}$ is the Fiedler eigenvector of $\Delta = D - A$. So we have reduced the problem to obtaining the Feidler eigenvector by standard means (the SVD for example).

Of course in this methodology we have only gone part of the way to clustering. We should look at the permuted similarity matrix $\hat{Q}$, where $\hat{Q}_{q_i, q_j} = Q_{i,j}$, and extract the blocks on the diagonals as clusters. This is a good place to start an argument between possible end-users.

Let us briefly return to the case where A is a binary adjacency matrix for an undirected network in A. In that case we had $g(z; f) = \delta(z)(1-f)+f\delta(z-1)$ and $\theta = f$, so that g yielded binary values for z with mean (probability that $z = 1$) equal to f. Now instead of using (1.2), as previously, we shall assume f to be range-k dependent, and simply set

$$f(k) = \frac{\gamma e^{-\eta k^2}}{1 + \gamma e^{-\eta k^2}}, \qquad (1.4)$$

where $\eta > 0$ and $\gamma > 0$ are constants. Therefore $f(k)$ in (1.4) is in (0,1) and decreasing and asymptotic to zero for large k.

Assuming each edge is independent of others, we generate such adjacency matrix by visiting each edge ($j > i$), and include it or not with the desired range-dependent probability. Then we impose symmetry and a zero main diagonal to complete the adjacency matrix.

Figure 1.6 shows an example of a range-dependent network, where A is binary.

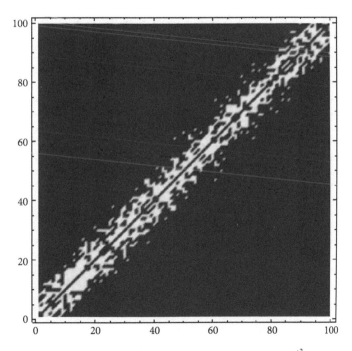

Fig 1.6 A range-dependent random network, given by $P(k) = \frac{\gamma_e^{-\eta k^2}}{1+\gamma e^{-\eta k^2}}$ with odds $f(k)/(1 - f(k)) = 0.8\,(0.98)^{k^2}$.

Hence, we have defined a random adjacency matrix for an undirected network. As we observed before, the expected value of our adjacency matrix is a Toeplitz matrix, with all of the elements on the kth super-diagonal (and by symmetry on the kth sub-diagonal) equal to $f(k)$. The random adjacency matrix is equivalent to a probability distribution defined over all possible adjacency matrices, $\mathcal{A}$. Formally $\mathcal{A}$ is the set of all symmetric matrices containing binary values, with a zero main diagonal: there are $2^{n(n-1)/2}$ elements in that set. Moreover, since the edges of our random undirected adjacency matrix are independent (by assumption), then for any candidate $R \in \mathcal{A}$, we may write down the probability distribution

$$P(R|X) = \prod_{i=1,\, j=i+1}^{n-1,\, n} f(j-i)^{R_{ij}}(1-f(j-i))^{(1-R_{ij})}.$$

This form is particularly elegant since there are exactly $n(n-2)/2$ terms in the product with each R_{ij} picking out the desired probability from the corresponding pair. Similar forms are used to derive maximum likelihood equations for logistic regression methods. This is a very useful construct indeed when dealing with binary variables (such as the R_{ij} here).

Now suppose that $\mathbf{q}$ is a permutation of the given network A, *then we have the* conditional 'likelihood' that the permuted version of A is generated by our random matrix:

$$\mathcal{L} \equiv P(A|\mathbf{q},\, X) = \prod_{i=1,\, j=i+1}^{n-1,\, n} f(|q_j - q_i|)^{A_{ij}}(1-f(|q_j - q_i|))^{(1-A_{ij})}.$$

But $f(k)$ in (1.4) may be rewritten as

$$\mathcal{L} \equiv P(A|\mathbf{q},\, X) = \prod_{i=1,\, j=i+1}^{n-1,\, n} f(|q_j - q_i|)^{A_{ij}}(1-f(|q_j - q_i|))^{(1-A_{ij})}.$$

we then rewrite this as follows:

$$\mathcal{L} \equiv P(A|\mathbf{q},\, X) = \left(\prod_{i=1,\, j=i+1}^{n-1,\, n} \left(\frac{f(|q_j - q_i|)}{(1-f(|q_j - q_i|))}\right)^{A_{ij}}\right)\left(\prod_{i=1,\, j=i+1}^{n-1,\, n} (1-f(|q_j - q_i|))\right).$$

The second product over all possible edges is independent of $\mathbf{q}$, while the first product is a product of odds for the observed edges. So to maximize $\mathcal{L}$ we may simply maximize the first product (over the edges that are observed).

Now from the choice of f in (1.4) [73], we have a simple form for the odds:

$$\frac{f(k)}{1 - f(k)} = \gamma e^{-\eta k^2}.$$

Taking logs (of the first product in our expression for $\mathcal{L}$) we have

$$\log \mathcal{L} = \frac{n(n-1)}{2} \log \gamma - \left(\sum_{i<j} A_{ij} \eta (q_i - q_j)^2 \right) + \text{constant}.$$

So, just as for the similarity matrix case, we should minimize $\log \tilde{\mathcal{L}}$ given by

$$\log \tilde{\mathcal{L}} = \sum_{i<j} A_{ij}(q_i - q_j)^2 = \frac{1}{2} \sum_{i,j=1}^{n} A_{ij}(q_i - q_j)^2 = \mathbf{q}^T.\Delta.\mathbf{q}.$$

Again we shall impose $\|\mathbf{q}\| = 1$ and $\mathbf{q.s} = 0$, since we only require the relative differences between the values in $\mathbf{q}$ in order to permute A.

Then everything else proceeds as before for matrices in $\mathcal{S}$, and we should exploit the elements of the Fiedler eigenvector so as to order the vertices (indices). Then the permuted version of A will have edges as close as possible to the main diagonal and will identify near cliques as blocks along the diagonal.

Of course the particularly simple form of this final minimization is made available by our choice of $f(k)$ in (1.4). Even when we wish to choose other definitions of $f(k)$ (for modelling reasons), it is usually possible to find parameters in (1.4) that produce a nearby distribution, especially for sparse matrices where we will need f to decay rapidly, so only the first few terms matter. Then even if we have to approximate a pre-desired f with one defined by (1.4), we will produce a near optimal reordering. The advantage of the Fiedler eigenvector as a solution is its availability (and very efficient calculation), via standard routines (including the SVD of A). The elegant choice made in (1.4) for a random binary (adjacency) matrix [73] is completely analogous to that made in (1.2) for the real valued similarity matrices.

It is very instructive to (a) generate an adjacency matrix from our given random adjacency matrix; and (b) calculate its Fiedler eigenvector (without bothering to shuffle it up). Then since it is not permuted we expect that the elements of the Fiedler eigenvector should almost always be ordered in monotonically increasing (or decreasing) fashion along the vector.

There is a very important point to be made here about the robustness of these ideas to observed errors, both false positives (edges observed in A that should not be there) and false negatives (edges not observed in A that should have been there). Given such an error-prone A, the ordering of the vertices produced by the Fielder eigenvector will eventually break down as errors increase. So one may test the sensitivity of a solution directly by randomly perturbing the optimized permuted version of A until the monotonicity of the Fiedler eigenvector is destroyed.

Depending upon the size of A we can use the SVD as a means of identifying the Fiedler eigenvector of a given non-negative similarity matrix or an undirected adjacency matrix A. Remember if zero is a double eigenvalue then it means that A is reducible into disconnected components. Once we have $\mathbf{q}$, we may reorder A so as to put it into the form most likely to have been generated by a range-dependent process, with subsets of indices that are heavily interrelated placed together. This is the precursor to a kind of investigation based on 'guilt by association': if we know some functions or attributes associated with some of the objects, it is reasonable to infer that other objects (permuted to be close by) may share those functions or attributes.

An Example From a Social Network

In Figure 1.7 is an observed adjacency matrix, A, from a network of pairwise friendships between individuals in a social network. There are $n = 121$ vertices (individuals). There are 542 given edges out of 7,260 possible edges. Taking the spectrum of the graph Laplacian ($\Delta = D - A$), we may see that in fact this network is completely connected (zero is a simple eigenvalue for Δ). In Figure 1.8 we show the permuted version of the adjacency matrix using the elements of the Fiedler eigenvector so as to reorder the indices/vertices of the network.

1.7 The Proteome, the Genome, and Beyond

Here we illustrate the methods covered in Section 1.6 with applications originally published in 2008 [55].

Since the time of Gregor Mendel, biologists have been attempting to understand how genes determine biological properties. How the 'genotype' (the organism's full set of genetic information) relates to the 'phenotype' (the organism's features and functionalities). Differences in genes largely explain biological diversity.

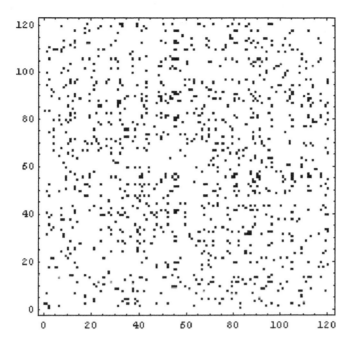

Fig 1.7 The raw social media data with vertices in some non-optimal order.

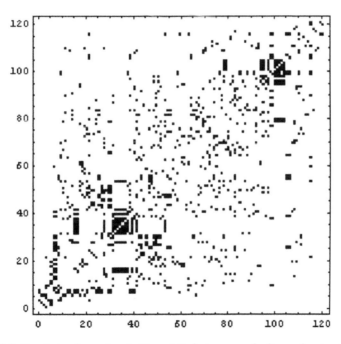

Fig 1.8 Reordering the vertices in Figure 1.7, placing near sub-cliques close together.

DNA forms an organism's genetic signature and may be viewed as a linear string where each character is one of the four nucleotide bases: C,A,T,G. It is arranged as a number of one-dimensional lattices (chromosomes). Certain contiguous chunks of DNA, that satisfy known constraints, may code for genes. Genes are important because they in turn code for proteins. Proteins are linear strings of amino acids, from an alphabet of twenty characters, but, unlike DNA, these strings fold into complicated 3D shapes, capable of interacting with each other in a myriad of ways. There are many references available for those who wish to learn more about basic cell biology from a mathematics/informatics perspective [17, 85].

Proteins are three-dimensional objects, and if two proteins are said to interact this means that they can physically combine. Experiments can be conducted where every possible pair of proteins in the cell can be tested to see if a mutual interaction takes place. The resulting protein–protein interaction (PPI) network is simply an undirected graph whose nodes are proteins and the edges denote observed interactions [53]. The emergence of such data (admittedly often containing both false-positives and false-negatives) raised a number of intriguing network-theoretic questions: how should such networks be characterized? How did they evolve? Where (relative to the whole) are the evolutionarily old and young proteins located? And do they contain small subnetworks of proteins that work together to produce common ends (cellular functions)?

Yeast two-hybrid (Y2H) experiments allow biologists to measure, in a pairwise fashion, whether proteins interact. The two-hybrid system is based on the premise that many eukaryotic transcriptional activators consist of two physically discrete modular domains. The DNA binding domain of the transcription factor is expressed as a hybrid protein fused to protein $P1$ (the 'bait'), while the activation domain is fused to protein $P2$ (the 'prey'). The domains act as independent modules: neither alone can activate transcription. Only if proteins $P1$ and $P2$ interact will the activation domain be in the proper position to activate transcription of the reporter gene. PPI networks obtained this way are very noisy; experimental limitations are believed to result in a rate of at least fifty percent for both the false negative (missing interactions) and false positive (spurious interactions) rates [52, 120]. In Figure 1.9 we plot the adjacency matrix of a PPI network for yeast based on the data from Uetz *et al.* [121]. Here, a dot in row i and column j indicates an interaction between proteins i and j. In this case there are 1048 proteins and 1029 interactions.

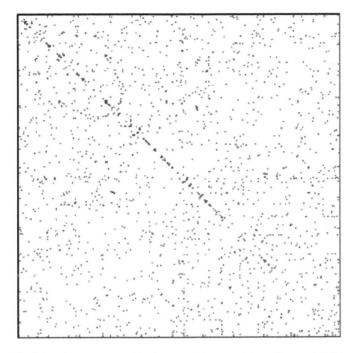

Fig 1.9 Adjacency matrix for the yeast PPI network from Uetz et al. [121].
A dot indicates a nonzero.

The PPI data take the form of an undirected (symmetric) adjacency matrix, $A \in \mathcal{A}$, where n is the number of proteins, with $A_{ij} = 1$ if proteins i and j are observed to interact and $A_{ij} = 0$ if there is no interaction.

In Figure 1.10 we show how the PPI network from Figure 1.9 looks when it is reordered according to the Fiedler vector, $\mathbf{q}$, determined via the SVD. In linear algebra terms, we have applied a symmetric row/column permutation to the adjacency matrix. We see that the network is made up largely of local interactions with relatively few long-range links, as proposed in range-dependent networks [52].

Next, let us consider gene expression data that are measured by microarrays. A single gene when active creates a single type of mRNA and in turn one type of protein. A microarray experiment records the activity of each set of individual genes simultaneously by measuring the levels (on a log scale) of their expression of the corresponding set of mRNA types. These data can be represented by a one-dimensional vector whose ith entry contains the expression level of the ith gene. Typically, data from several experiments will be collected.

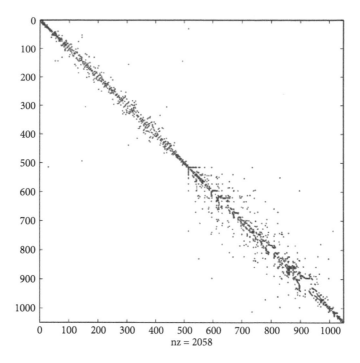

Fig 1.10 The adjacency matrix for the PPI network in Figure 1.9 when reordered using the Fiedler eigenvector.

For example, tissue from different cancer patients may be tested, or a single tissue may be tested at different times in order to produce a time series. In both cases, the different experiments are usually referred to as samples, and the resulting data set can be thought of as a two-dimensional array, with the jth column representing the one-dimensional output for the jth sample.

These mRNA measurements are often called gene-expression data and they give a snapshot of the state of transcription of each sample, which in turn reflects the relative importance of their proteins within that tissue. Some leukaemia patients, for example, overproduce red blood cell mRNA and proteins including haemoglobin, responsible for the red colour of blood and oxygen transport. On the other hand, the normal genes for lung function, including the surfactant proteins, which prevent the lungs from collapsing, are often switched off in lung cancer tumours (the tumour is usually solid and has no use for the surfactant). Hence repressed activity levels of those proteins are possible indicators for that disease.

We shall assume that the microarray data take the form of a non-negative $n \times m$ matrix, W, where n is the number of genes (rows) and m is the number of samples (columns), with $W_{ij} \geq 0$ measuring the activity (expression level) of gene i in sample j. The larger the value W_{ij}, the greater the activity level. We have in mind the case $n > m$ (many genes and few samples) though our ideas and analysis hold equally well for $m \geq n$.

This situation is entirely analogous to a whole range of analytics applications in other analytics fields such as customer, social, and societal applications. For example, we may monitor n people with respect to m conditions, or at m distinct moments in time, and W_{ij} might be a (crude) non-negative measure of the economic, social, or even political activity level of the ith person under the jth condition. For example, we may observe millions (n) of shoppers' purchasing behaviour in response to a set of m distinct promotions of various types (buy-one-get-one-free, etc. off, fifty percent) across many product categories, with promotions within fresh categories leading to increased immediate consumption, and thus being capped; while promotions within ambient or household categories may lead to stockpiling, cannibalizing future purchases. Or else we may monitor individuals' activity on social media platforms in response to distinct types of news alerts (conflict, terrorism, banking crises, sports, etc).

We shall take, as similarity measures, the elements in the $n \times n$ symmetric similarity matrix $A = WW^T - D'$ where D' is the diagonal matrix containing the diagonal element from WW^T. Strictly speaking we should make sure that elements of A lie on $[0,1]$ so that $A \in \mathcal{S}$.

In Solutions to exercise section we discuss the theoretical construct of such *activity* matrices, derived for n consumer products based on their presence within m ($>> n$) shopping baskets, so that W may well be too big to hold, yet A, though large, may be manageable. The SVD of WW^T can be related to that of W, so we may apply the SVD to whichever is the smaller (where $n < m$ and $m < n$, respectively).

The projection of the n elements (genes, customers, etc.) under consideration onto the real valued measures given by the corresponding elements of the Fiedler eigenvector (the second eigenvector of the Laplacian) should separate them out according to their similar or dissimilar performance within the samples. It is often the case that other rearrangements of the elements are also relevant, and arguments along the lines of those above can be used to show that the third, fourth, etc., eigenvectors of Δ are natural candidates. One way to justify this generalization is developed in a paper by Higham, Kalna and Kibble (2007) [75].

To illustrate the performance of the SVD, we present some results on microarray data from cancer studies. Here each sample, corresponding to each column of W, is from the tumour of a patient with a known type of cancer, or from a normal/control tissue. In these examples we are simply testing whether the SVD can rediscover the known groupings, but it should be clear that a successful algorithm has enormous potential for revealing new information and answering questions like those above.

Figure 1.11 gives the results. In each of the four cases we have used both the second (cf. Fiedler) and third singular vectors to give two-dimensional components for the samples, producing a Òscatter plot where nearby samples are likely to be related. Let us re-emphasize that the underlying idea here is to use correlation in gene expression behaviour in order to classify sample types. In the upper left scatter plot, clear separation of the carcinoma (stars) and normal (circles) kidney samples has been achieved by the second singular vector in data from Choi et al. (2005) [24]. The upper right part of Figure 1.11 shows a scatter plot of the second versus third eigenvectors of seventy-two leukaemia

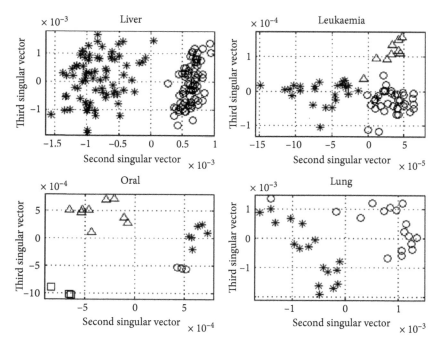

Fig 1.11 Upper left: carcinoma (stars) and normal (circles) liver samples; here $m=12,065$ and $n=152$. Upper right: T (triangles) and B (circles) cell ALLs, AML (stars); here $m=7129$ and $m=72$. Lower left: normal in vivo (squares) and in vitro (circles) and carcinomas in vivo (triangles) and in vitro (stars); here $m=20,428$ and $n=21$. Lower right: carcinoma (stars) and normal (circles) lung samples; here $M=5,983$ and $N=34$.

samples. In this plot the samples are known to divide into three different groups. We see that the second singular vector does a good job of separating the ALLs (acute lymphoblastic leukaemia) and AMLs (acute myeloid leukaemia), while the third singular vector focuses on distinguishing between the T and B subtypes of ALL. Four clear subgroups can be seen in the lower left part of Figure 1.11. The second, dominant, singular vector separates in vivo and in vitro samples and the third vector divides normal samples from oral carcinomas. These data come from the first set of experiments of an undergoing study led by Thurlow at the Beatson Institute for Cancer Research, Glasgow.

Along with clustering/ordering information, it is possible to compute measures of sensitivity, giving information about how reliable the results are in the presence of uncertainty in the data [117].

The expression levels in microarray data can take the form of signed data, with negative values representing under-expression [76]. Matrix-based inference algorithms can be developed through a systematic Bayesian approach [98], and in the case where data represent evolution, algorithms for time series may be appropriate [44]. As an alternative to the SVD, a non-negative matrix factorization can be used [45]. Expression data from two different organisms can be analysed with the Generalized Singular Value Decomposition [3].

A Personal View

On Similarity Matrices and Basket Metrics

One of the first times that I encountered very large similarity matrices was in analysing basket data from supermarkets in the mid-to-late 1990s, when loyalty card data became available. These retailers sell maybe 10^5 distinct product items, called stock-keeping units (SKUs). We think of the sales of units of each product of interest, $i = 1, \ldots, n$, sold over a given time interval, as a distribution over the (very large) set of baskets (or over the set customer IDs) of some size $m \gg n$. Then we wish to measure the similarity of those distributions for all possible SKU pairs, (i, j). We may do this by normalizing the sales of each SKU over all baskets (or all customers) so as to sum to unity, and then taking a dot product between pairs of those normalized distributions. Conceptually this is a running sum over all baskets (or customers) of the product of the corresponding normalized sales of each SKU. Since it is the dot product of two non-negative vectors on the standard m-simplex in $\mathbb{R}^m$, this product is always in $[0,1]$, and it is equal to one if, and only if, both SKUs are only sold exclusively to one and the same basket/customer. The $n \times n$ similarity

continued

On Similarity Matrices and Basket Metrics *Continued*

On matrix obtained is thus in S, and it may be achieved by winding through the m baskets (or customers) one at a time, and keeping track of total sales of each item, so it is not necessary to hold all of the data in memory. This method can apply equally to sets of product items defined at any level within a product hierarchy, from sets of categories down to subcategories and SKUs. So i might be 'any bottle of navy rum', and j might be 'any packet of ibuprofen tablets'. (Note that beer and are nappers more common examples.)

One soon found out that hair colourant is not sold with shampoo or hair conditioner at all, where they are often ranged, but is instead cross-sold with make-up and jewellery items; and that various cross-category combinations could be usefully co-located or cross-promoted. In the late nineties this was counter to the usual category-based way of managing products.

One way to present the insights from pulling sets of cross-purchased, niche SKUs together was in a dendrogram. We successively combined the product pair (i,j) with the largest pairwise similarity so that they played-on as a joint product, 'i and/or j'. This is like an enhanced version of the FA cup, where a winning team keeps both the players and supporters of the team that they vanquish as they progress to the next rounds. In combining products in this way we took the appropriate (total sales) weighted sum of the ith and jth rows (and columns) to form up the new combined row (column) in the similarity matrix. Hence it was not necessary to wind through the entire m-baskets/customers data again in order to update the similarity matrix. Successively one combined the products, bottom up, by operating directly on that $n \times n$ similarity matrix. This was highly efficient. By successively combining the most niche SKUs together (within a dendrogram) it provided a view across the whole range at whatever hierarchy level that was desired, see Figure 1.12 for example. But that was only the start. We required a further level of analytics to extract further value.

We achieved this by identifying relevant sets, or portfolios, of cross-purchased products (usually across categories) from the above similarity matrix, and then by deriving their pairwise conditional probabilities of purchasing that a basket would contain product j given that it contained product i. Some products are highly likely to be in the basket given that the basket contains almost anything else from within the portfolio in question. These are termed 'pivotal products': once a customer is in the mode to buy from that portfolio, they will also tend to place the pivotal products into their basket. Other products act as triggers: if they are present within a basket then almost every other product within the portfolio is much more likely to be also present. These are termed 'gateway' or 'trigger' products. (Note that the causal language is not rigorously justified.) Pivotals act to drive

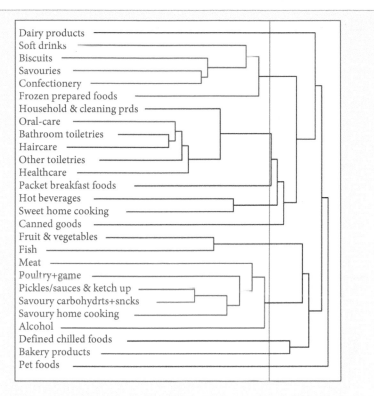

Fig 1.12 Dendrogram based on combining subcategories within the similarity matrix.

traffic, in the store or online, while triggers act to build the basket respectively, by pulling the customer into a mode that is focussed on the set in question. They need to be ranged and promoted appropriately, using pivotals to reward both loyalty and ubiquitous purchasing, and using triggers to shift customers into new modes of shopping that they may otherwise be giving to other retailers. For example in a component meal portfolio, frozen vegetables are pivotals while meat and fish products are triggers; in a baby-shopping portfolio nappres are pivotals, while bibs and changing mats are triggers.

Armed with such insights, data-rich retailers gave up their old supply-centric approach (category and supply chain management, striving to put SKUs onto their shelves efficiently), and instead became customer-centric (understanding why customers might take SKUs off the shelves). Online recommender systems may exploit the conditional probabilities and the pivotals, while in-store and online substitute SKUs for out-of-stock SKUs can be auto-generated by finding products that are heavily cross-purchased over time by loyalty card or account holders but

continued

On Similarity Matrices and Basket Metrics *Continued*

that are hardly ever cross-purchased within the same basket (implying some mutual exclusivity with switching driven by choice, promotion, availability). This last is essential in working out the cost of stock outages. Some SKUs will have no substitute.

Basket metrics are now very standard within retailers with loyalty data and/or disaggregated basket data.

On Graphs with Structure

On our summer holiday in Galicia, I first read the book *Small Worlds* by Duncan J. Watts [125]. This inspired an interest in range-dependent networks, the inverse problem described in this chapter, and the behaviour of the Watts–Strogatz clustering coefficient for such networks as parameters varied. This would have remained unremarked were it not for Steven Strogatz who encouraged me to publish when I emailed him, which illustrates the importance of people showing an interest in each other's ideas. In turn this led to work on PPI networks and the development of academic collaborations to research into networks and matrices involving the SVD and other ideas during the early 2000s. This was an exciting time, especially as we sought to group entities (genes and proteins) of unknown function with similar entities of known function(s). Many of the concepts and methods I had been applying to customer behaviour for retailers' loyalty card data were useful within the analysis of 'omic data, and vice versa. Back then, possibly too much bioinformatics research was focussed on data handling, normalization, and curation: there was a lack of analytics and a large waste of public money in the setting up of web-accessible databases.

Graphs with structure are now big business. A motif is a certain subgraph (usually relatively small) that may occur within any network under examination, and it is a popular pastime to analyse the distribution of different motifs within certain classes of graphs. Yet within any application 'the data are the data', so how can this provide actionable insight? It certainly produces signature properties of a given network and allows for the inter-comparison of networks, but an end-user would remain unconvinced unless they can show that such insight yields some advantages. Do we analysts challenge ourselves to do so?

CHAPTER 2

Dynamically evolving networks

Suppose that we need to observe a network of people that represents their social interactions or their regular or irregular communications. Then it is reasonable to seek out the most important people, those with the most potential to influence others. There are a number of ways to define a real valued measure that indicates some aspect of the individuals' relative importance, or the influence of individuals: these are called *centrality measures* (for example, see [12]). For a fixed (static) graph, such as a mutual friendship network (that hopefully only changes on a long timescale), we might propose degree centrality, where the vertex degree is the measure. Or else we may count the number of shortest paths between any pairs of vertices that happen to flow through a particular vertex.

In this chapter we are interested in defining a centrality measure that acknowledges that a vertex does not need to have a large degree in order to be to be influential. If you know people who know people who know people, you may be more effective and efficient at disseminating and picking up information rather than if you were connected to a large number of relative loners. Yet we also wish to take an efficient (and algebraic) approach rather than doing some costly path-following and path-versus-path comparisons. Moreover we wish to extend the idea of centrality to work for dynamically evolving networks, so that it can measure the instances of time-ordered cascades of peer-to-peer communications. This is a situation we shall encounter often as news, rumours,

Mathematical Underpinnings of Analytics. First Edition. Peter Grindrod.
© Peter Grindrod 2015. Published in 2015 by Oxford University Press.

information, attitudes, ideas, social norms, and other entities are propagated within social communications networks.

Here we shall focus on deriving variants of the so-called Katz centrality, which is in turn related to an alternative construct called eigenvector centrality (given by the elements of the Perron–Frobenius eigenvector of the adjacency matrix) and to Google's page rank. We shall see below that it is based on the matrix function $(I - \alpha A)^{-1}$, where A is an adjacency matrix and α is a small enough positive discounting constant. The ideas given here though will also extend to other forms of centrality (for example those based on the matrix function $(I + A)$ or on the matrix exponential function $\exp(A)$, alternatives which have been proposed in the past, see [40]).

2.1 The Katz Centrality Matrix for a Static Network

Suppose we observe a network of friends forming a social network on the internet. For a variety of marketing, targeting, and monitoring applications we may want to know which people are well-connected and which are most influential. For example, we may wish to start a rumour, or to distribute some propaganda or marketing messages. Such information may be propagated from friend to friend: who should we influence or communicate with to ensure the information is spread as wisely as possible? Alternatively we may wish to find out what is going on: who should we ask? Intuitively you might seek the person (vertex) with the largest degree. But if that person is a local 'hub', possessing many adjacencies that are vertices of degree one, then the message may not get very far. So we need to find a candidate vertex that can efficiently reach the rest of the network with short paths, possibly via many alternative routes. Perhaps there are people who are low degree hubs of larger hubs.

Let $\mathcal{A}$ and $\mathcal{S}$ be as in Chapter 1. So $A \in \mathcal{A}$ is an adjacency matrix for an undirected graph.

Consider the vertex i; how may it propagate information along walks (successive vertex-to-vertex edge connections) to vertex j? We know there is walk of length one from i to j if $(A)_{ij} = 1$, and (by a theorem given in Chapter 1) $(A^2)_{ij}$ counts the walks of length two. Similarly $(A^k)_{ij}$ counts the possible walks of length k. We want to count all such walks, since each represents a possible route for the information from vertex i to vertex j. Suppose the information has a probability α of being successfully transmitted between any adjacent pair of vertices (let us assume this acts independently for each edge each time the information is transferred). Then it makes sense to discount the walks of length k by a factor α^k. This is simply the probability that the transmission is successful. Alternatively we may simply wish pragmatically to discount long range edges as being less important.

Consider the matrix

$$Q = I + \alpha A + \alpha^2 A^2 + \alpha^3 A^3 + \cdots = (I - \alpha A)^{-1}.$$

Here, α must be small enough so that the geometric sum exists. To be precise we must have $\alpha \rho(A) < 1$, where $\rho(A)$ is the spectral radius (the Perron–Frobenius eigenvalue, see Chapter 1) of A.

Then assuming this is so, we see that for $i \neq j$ the sum held in Q_{ij} is precisely the term we require to describe the connectivity of vertex i to vertex j. This sum gives us a key performance measure of the ability of vertex i to communicate successfully with j.

Hence Q_{ij} denotes a measure of the possible flow of information from vertex i to vertex j.

We shall call Q the *centrality matrix* for A. It was first introduced by Katz [86].

Notice that for a non-negative symmetric matrix A, all of its powers and hence Q are also non-negative and symmetric. Let us consider the row sums and column sums of Q:

$$\mathbf{b} = Q\mathbf{s} \quad \mathbf{r} = Q^T\mathbf{s},$$

where $\mathbf{s} = (1, 1, \ldots, 1)^T \in \mathbb{R}^n$. These are equal since Q is symmetric, but we shall denote these separately as $\mathbf{b}$ for 'broadcast' and $\mathbf{r}$ for 'receive' (as we shall shortly encounter situations where they are not equal).

The ith elements of $\mathbf{b}$ and $\mathbf{r}$ give a very straightforward measure of how effectively the vertex i can reach every other vertex and be reached from every other vertex.

So we should plant our information (or our listening device) at the vertex which enjoys the highest such row/column sum. If we have only one option, this identifies the best vertex to communicate to (start a rumour) and from (hear a rumour) all of the other vertices.

The *resolvent* function $(I - \alpha A)^{-1}$ in the definition of Q is an example of a matrix values function [78] (see Section 2.2). In this case it is very straightforward since A is normal (real and symmetric) and thus so is $(I - \alpha A)$.

The condition that $\alpha\rho(A) < 1$ is enough to ensure that the inverse of $(I - \alpha A)$ exists. Indeed it is clear the geometric series in the definition of $\mathcal{Q}$ converges in the matrix norm.

For example consider the graph in Figure 2.1 with adjacency matrix

$$A = \begin{pmatrix}
0 & 1 & 0 & 0 & 1 & 1 & 0 & 0 & 0 & 0 & 0 & 0 & 0 & 0 & 0 & 1 & 1 & 0 & 0 & 1 \\
1 & 0 & 1 & 1 & 1 & 1 & 0 & 0 & 0 & 1 & 1 & 0 & 0 & 0 & 0 & 0 & 1 & 1 & 0 & 0 \\
0 & 1 & 0 & 0 & 1 & 0 & 1 & 1 & 1 & 0 & 0 & 0 & 0 & 1 & 1 & 1 & 0 & 0 & 0 & 0 \\
0 & 1 & 0 & 0 & 0 & 1 & 1 & 1 & 0 & 1 & 0 & 1 & 0 & 0 & 0 & 1 & 0 & 0 & 0 & 0 \\
1 & 1 & 1 & 0 & 0 & 1 & 1 & 1 & 1 & 0 & 0 & 0 & 0 & 0 & 0 & 0 & 0 & 0 & 0 & 0 \\
1 & 1 & 0 & 1 & 1 & 0 & 1 & 0 & 1 & 0 & 0 & 0 & 1 & 1 & 1 & 0 & 0 & 1 & 1 & 0 \\
0 & 0 & 1 & 1 & 1 & 1 & 0 & 0 & 1 & 1 & 0 & 1 & 0 & 1 & 0 & 0 & 0 & 0 & 1 & 0 \\
0 & 0 & 1 & 1 & 1 & 0 & 0 & 0 & 0 & 1 & 1 & 1 & 0 & 1 & 0 & 0 & 0 & 0 & 1 & 0 \\
0 & 0 & 1 & 0 & 1 & 1 & 1 & 0 & 0 & 1 & 0 & 1 & 0 & 0 & 1 & 0 & 0 & 0 & 0 & 0 \\
0 & 1 & 0 & 1 & 0 & 0 & 1 & 1 & 1 & 0 & 0 & 0 & 1 & 1 & 0 & 1 & 1 & 0 & 0 & 0 \\
0 & 1 & 0 & 0 & 0 & 0 & 0 & 1 & 0 & 0 & 0 & 1 & 0 & 1 & 1 & 0 & 1 & 0 & 1 & 0 \\
0 & 0 & 0 & 1 & 0 & 0 & 1 & 1 & 1 & 0 & 1 & 0 & 1 & 0 & 0 & 0 & 0 & 0 & 1 & 0 \\
0 & 0 & 0 & 0 & 0 & 1 & 0 & 0 & 0 & 1 & 0 & 1 & 0 & 1 & 1 & 1 & 1 & 0 & 1 & 1 \\
0 & 0 & 1 & 0 & 0 & 1 & 1 & 1 & 0 & 1 & 1 & 0 & 1 & 0 & 0 & 1 & 1 & 1 & 1 & 0 \\
0 & 0 & 1 & 0 & 0 & 1 & 0 & 0 & 1 & 0 & 1 & 0 & 1 & 0 & 0 & 0 & 0 & 0 & 1 & 0 \\
1 & 0 & 1 & 1 & 0 & 0 & 0 & 0 & 0 & 1 & 0 & 0 & 1 & 1 & 0 & 0 & 0 & 1 & 1 & 1 \\
1 & 1 & 0 & 0 & 0 & 0 & 0 & 0 & 0 & 1 & 1 & 0 & 1 & 1 & 0 & 0 & 0 & 0 & 1 & 0 \\
0 & 1 & 0 & 0 & 0 & 1 & 0 & 0 & 0 & 0 & 0 & 0 & 0 & 1 & 0 & 1 & 0 & 0 & 0 & 0 \\
0 & 0 & 0 & 0 & 0 & 1 & 1 & 1 & 0 & 0 & 1 & 1 & 1 & 1 & 1 & 1 & 1 & 0 & 0 & 1 \\
1 & 0 & 0 & 0 & 0 & 0 & 0 & 0 & 0 & 0 & 0 & 0 & 1 & 0 & 0 & 1 & 0 & 0 & 1 & 0
\end{pmatrix}.$$

Then we have $\rho(A) = 8.26\ldots$. So let us set $\alpha = 0.1$. Then we have $Q =$

$$\begin{pmatrix}
1.13 & 0.24 & 0.13 & 0.12 & 0.21 & 0.27 & 0.14 & 0.12 & 0.11 & 0.15 & 0.11 & 0.09 & 0.16 & 0.18 & 0.09 & 0.23 & 0.21 & 0.09 & 0.18 & 0.17 \\
0.24 & 1.24 & 0.27 & 0.27 & 0.27 & 0.35 & 0.23 & 0.2 & 0.17 & 0.31 & 0.25 & 0.16 & 0.2 & 0.28 & 0.15 & 0.21 & 0.28 & 0.21 & 0.24 & 0.09 \\
0.13 & 0.27 & 1.23 & 0.18 & 0.28 & 0.26 & 0.32 & 0.28 & 0.27 & 0.23 & 0.17 & 0.17 & 0.2 & 0.35 & 0.24 & 0.28 & 0.16 & 0.11 & 0.25 & 0.08 \\
0.12 & 0.27 & 0.18 & 1.2 & 0.16 & 0.31 & 0.31 & 0.27 & 0.16 & 0.31 & 0.15 & 0.25 & 0.2 & 0.25 & 0.12 & 0.27 & 0.15 & 0.11 & 0.24 & 0.08 \\
0.21 & 0.27 & 0.28 & 0.16 & 1.19 & 0.32 & 0.25 & 0.25 & 0.19 & 0.13 & 0.14 & 0.16 & 0.22 & 0.13 & 0.16 & 0.14 & 0.09 & 0.2 & 0.07 \\
0.27 & 0.35 & 0.26 & 0.31 & 0.32 & 1.37 & 0.39 & 0.25 & 0.31 & 0.29 & 0.22 & 0.23 & 0.37 & 0.43 & 0.29 & 0.27 & 0.23 & 0.24 & 0.42 & 0.13 \\
0.14 & 0.23 & 0.32 & 0.31 & 0.29 & 0.39 & 1.31 & 0.25 & 0.31 & 0.35 & 0.19 & 0.3 & 0.25 & 0.4 & 0.18 & 0.24 & 0.19 & 0.12 & 0.38 & 0.1 \\
0.12 & 0.2 & 0.28 & 0.27 & 0.25 & 0.25 & 0.25 & 1.24 & 0.18 & 0.31 & 0.28 & 0.28 & 0.22 & 0.37 & 0.15 & 0.21 & 0.18 & 0.1 & 0.35 & 0.09 \\
0.11 & 0.17 & 0.27 & 0.16 & 0.25 & 0.31 & 0.31 & 0.18 & 1.19 & 0.27 & 0.14 & 0.24 & 0.18 & 0.23 & 0.23 & 0.16 & 0.13 & 0.08 & 0.22 & 0.06 \\
0.15 & 0.31 & 0.23 & 0.31 & 0.19 & 0.29 & 0.35 & 0.31 & 0.27 & 1.29 & 0.2 & 0.21 & 0.34 & 0.41 & 0.16 & 0.33 & 0.3 & 0.13 & 0.3 & 0.11 \\
0.11 & 0.25 & 0.17 & 0.15 & 0.13 & 0.22 & 0.19 & 0.28 & 0.14 & 0.2 & 1.19 & 0.25 & 0.21 & 0.33 & 0.22 & 0.17 & 0.26 & 0.09 & 0.34 & 0.08 \\
0.09 & 0.16 & 0.17 & 0.25 & 0.14 & 0.23 & 0.3 & 0.28 & 0.24 & 0.21 & 0.25 & 1.19 & 0.27 & 0.24 & 0.15 & 0.17 & 0.15 & 0.08 & 0.33 & 0.08 \\
0.16 & 0.2 & 0.2 & 0.2 & 0.16 & 0.37 & 0.25 & 0.22 & 0.18 & 0.34 & 0.21 & 0.27 & 1.29 & 0.41 & 0.26 & 0.33 & 0.3 & 0.13 & 0.41 & 0.22 \\
0.18 & 0.28 & 0.35 & 0.25 & 0.22 & 0.43 & 0.4 & 0.37 & 0.23 & 0.41 & 0.33 & 0.24 & 0.41 & 1.42 & 0.22 & 0.39 & 0.35 & 0.25 & 0.47 & 0.14 \\
0.09 & 0.15 & 0.24 & 0.12 & 0.13 & 0.29 & 0.18 & 0.15 & 0.23 & 0.16 & 0.22 & 0.15 & 0.26 & 0.22 & 1.15 & 0.15 & 0.14 & 0.08 & 0.3 & 0.08 \\
0.23 & 0.21 & 0.28 & 0.27 & 0.16 & 0.27 & 0.24 & 0.21 & 0.16 & 0.33 & 0.17 & 0.17 & 0.33 & 0.39 & 0.15 & 1.26 & 0.2 & 0.21 & 0.36 & 0.22 \\
0.21 & 0.28 & 0.16 & 0.15 & 0.14 & 0.23 & 0.19 & 0.18 & 0.13 & 0.3 & 0.26 & 0.15 & 0.3 & 0.35 & 0.14 & 0.2 & 1.2 & 0.1 & 0.33 & 0.1 \\
0.09 & 0.21 & 0.11 & 0.11 & 0.09 & 0.24 & 0.12 & 0.1 & 0.08 & 0.13 & 0.09 & 0.08 & 0.13 & 0.25 & 0.08 & 0.21 & 0.1 & 1.09 & 0.15 & 0.05 \\
0.18 & 0.24 & 0.25 & 0.24 & 0.2 & 0.42 & 0.38 & 0.35 & 0.22 & 0.3 & 0.34 & 0.33 & 0.41 & 0.47 & 0.3 & 0.36 & 0.33 & 0.15 & 1.39 & 0.23 \\
0.17 & 0.09 & 0.08 & 0.08 & 0.07 & 0.13 & 0.1 & 0.09 & 0.06 & 0.11 & 0.08 & 0.08 & 0.22 & 0.14 & 0.08 & 0.22 & 0.1 & 0.05 & 0.23 & 1.08
\end{pmatrix}.$$

The row sums of Q are given by (4.21999, 5.71866, 5.56521, 5.2072, 4.95972, 7.05118, 6.37333, 5.67824, 4.99056, 6.27106, 5.0908, 5.07467, 6.20914, 7.41627, 4.61043, 5.90485, 5.21234, 3.6091, 7.19744, 3.35314).

The vertex with the largest off-diagonal row sum is number 14 (see Figure 1.1). That is where we should listen to the communications or input our information to be propagated.

Exercise 2.1

The so called eigenvector centrality of a static network, with adjacency matrix A, is defined by the elements of the Perron–Frobenius eigenvector, $\mathbf{v}_{PF}$:

$$\lambda_{PF}\mathbf{v}_{PF} = A\mathbf{v}_{PF}.$$

Yet above we have $\mathbf{r} = (I - \alpha A)^{-1}\mathbf{s}$.

Let $\mathbf{w} = \mathbf{r} - \mathbf{s}$. Show that $\mathbf{w} = \alpha A(\mathbf{w} + \mathbf{s})$. Expand $\mathbf{w}$ and $\mathbf{s}$ as eigenvector expansions using the eigenvectors of the normal matrix A. What happens as $1/\alpha \rightarrow \lambda_{PF}$ from above?

2.2 Matrix Valued Functions

Before going any further, we pause to consider some properties of matrix valued functions that will be useful. In particular we wish to define fractional powers of matrices, as well as the matrix logarithm. A matrix multivalued, so we wish to take its primary value, while other functions, such as e^A or the resolvent $(I - \alpha A)^{-1}$ that we have met already in Chapter 1, simply need to be well-defined.

There are a number of ways to define matrix valued functions and we shall focus on the simplest. An excellent source of further reading is the book by Higham [78].

Suppose A in an $n \times n$ matrix with complex valued elements. Then it can be written in its Jordan canonical form:

$$A = ZJZ^{-1},$$

where Z is non-singular and

$$J = \text{diag}(J_1, J_2, \ldots, J_q)$$

is a block diagonal matrix where each J_k corresponds to an eigenvalue λ_k, has dimension $n_k \times n_k$, and is of the form

$$J_k = \begin{pmatrix} \lambda_k & 1 & 0 & \cdots & 0 \\ 0 & \lambda_k & 1 & \cdots & 0 \\ \vdots & \vdots & \vdots & \vdots & \vdots \\ 0 & \cdots & 0 & \lambda_k & 1 \\ 0 & \cdots & \cdots & 0 & \lambda_k \end{pmatrix}.$$

Clearly $n_1 + \cdots n_q = n$ and J is unique up to any reordering of the blocks J_k.

For real self-adjoint matrices (similarity matrices and adjacency matrices for undirected graphs) we have a unitary matrix Z ($Z^{-1} = Z^T$) and A is diagonalizable, so all of the blocks are of size $n_k = 1$, and thus J is diagonal.

Proceeding generally for any function f defined on some domain in $\mathbb{C}$, we say that f is defined on the spectrum of A if all of the values for f and its derivatives

$$f(\lambda_k), f^{(1)}(\lambda_k), \ldots, f^{(n_j-1)}(\lambda_k)$$

exist. In that case we define

$$f(A) = Zf(J)Z^{-1} = Z\mathrm{diag}(f(J_k))Z^{-1},$$

where

$$f(J_k) = \begin{pmatrix} f(\lambda_k) & f^{(1)}(\lambda_k) & \frac{f^{(2)}(\lambda_k)}{2!} & \cdots & \frac{f^{(n_k-1)}(\lambda_k)}{(n_k-1)!} \\ 0 & f(\lambda_k) & f^{(1)}(\lambda_k) & \cdots & \frac{f^{(n_k-2)}(\lambda_k)}{(n_k-2)!} \\ \vdots & \vdots & \vdots & \vdots & \vdots \\ 0 & \cdots & 0 & f(\lambda_k) & f^{(1)}(\lambda_k) \\ 0 & \cdots & \cdots & 0 & f(\lambda_k) \end{pmatrix}.$$

Usually we would have f given by some closed formula that may be evaluated on a large domain, but note that the definition only needs knowledge of f and its derivatives at the eigenvalues of A.

Where does this expression for $f(J_k)$ come from? First $J_k = \lambda_k I + N_k$ where

$$N_k = \begin{pmatrix} 0 & 1 & 0 & \cdots & 0 \\ 0 & 0 & 1 & \cdots & 0 \\ \vdots & \vdots & \vdots & \vdots & \vdots \\ 0 & \cdots & 0 & 0 & 1 \\ 0 & \cdots & \cdots & 0 & 0 \end{pmatrix}.$$

It is clear that

$$
N_k^2 = \begin{pmatrix}
0 & 0 & 1 & \cdots & 0 \\
\vdots & \vdots & \vdots & \vdots & \vdots \\
0 & 0 & 0 & \cdots & 1 \\
0 & \cdots & 0 & 0 & 0 \\
0 & \cdots & \cdots & 0 & 0
\end{pmatrix}.
$$

Each successive power of N_k moves the super diagonal of 1s one place further towards the top right-hand corner. Thus $N_k^{n_k} = 0$.

Now suppose we are given a function f defined on the spectrum of A, and thus each J_k, and we consider a Taylor series expansion for f:

$$
f(z) = f(\lambda_k) + f^{(1)}(\lambda_k)(z - \lambda_k) + \cdots + \frac{f^{(j)}(\lambda_k)}{j!}(z - \lambda_k)^j + \cdots
$$

Let us substitute $J_k = \lambda_k I + N_k$ for z. Then we obtain a finite series:

$$
f(J_k) = f(\lambda_k)I + f^{(1)}(\lambda_k)N_k + \cdots + \frac{f^{(j)}(\lambda_k)}{(n_k - 1)!}N_k^{n_k-1}.
$$

Hence our definition for $f(J_k)$ is given above.

We may define the integer powers of a matrix A, polynomials of A, and the exponential of A, corresponding exactly to our usual notions. Note that for two $n \times n$ matrices, A_1 and A_2, it is not generally true that $e^{A_1+A_2} = e^{A_1}e^{A_2}$. Indeed this is only true if these matrices commute, so some care must be taken.

We may take fractional powers by taking fractional power function $f(z)^{1/m}$, $m \in \mathbb{Z}^+$ defined consistently within each block in the normal form. Since there are q such blocks and m branches for the mth root, there are at least m^q matrices B such that $B^m = A$. These are called primary roots. In fact there are other, non-primary roots, if $q < n$ (meaning at least one Jordan block has $n_j > 1$), see [78].

The matrix logarithm, $\log(B)$, is well defined provided B has no eigenvalues that are real and negative. It yields matrices C such that $e^C = B$. Because of the multivalued nature of the complex logarithm, there are an infinite number of such matrices (where we add 2π integer multiples of imaginary blocks), yet there is a unique principal logarithm whose spectrum lies within the strip $|\arg(z)| < \pi$. For matrices B with real and positive eigenvalues things are straightforward: we simply employ the above definition using that real valued logarithm defined as $(0, \infty)$.

So in the case of the Katz centrality, we are interested in $f(z) = (1 - \alpha z)^{-1}$, which was well-defined so long as the spectral radius of A, $\rho(A)$, is less than $1/\alpha$. Now we can see this function is also meaningful (even though the geometric series representation does not converge) provided no eigenvalue is equal to $1/\alpha$.

Since we shall usually have A real and symmetric then all of its eigenvalues are real, and if all are less than $1/\alpha$, then $B = (I - \alpha A)$ has all real and positive eigenvalues (in $[1 - \rho(A)/\alpha, 1 + \rho(A)/\alpha]$). Thus the log of B is well-defined.

Next, for all $\alpha \in [-1, 1]$, if B has no eigenvalues on the negative real axis then we have (see [78])

$$\log(B^\alpha) = \alpha \log B.$$

So, for example, $\log(B^{-1}) = -\log B$, and all real powers of B less than one can follow.

Thus, with the exception of taking functions of sums of matrices that do not commute, the fractional powers, the logarithm function, and the real powers correspond to the scalar situation. Moreover the case when A, and thus $(I - \alpha A)$, is diagonalizable (which we shall usually encounter) is particularly straightforward: we may keep the Jordan decomposition and apply the function to the (real) eigenvalues in such cases.

Exercise 2.2

Let A be the adjacency matrix for an undirected graph. Then A is diagonalizable. How many square roots does A have (that is matrices B for which $B^2 = A$)?

Exercise 2.3

Let A be the adjacency matrix for an undirected graph. We write $A = U\Lambda U^T$, where U is a unitary matrix with columns given by eigenvectors of A, and $\Lambda = \text{diag}(\lambda_1, \lambda_2, \dots \lambda_n)$ is a real diagonal matrix of eigenvalues. Show that for all polynomials, P, of any degree there exists a polynomial R of degree less than n such that $P(A) = UR(\Lambda)U^T$, where $R(\Lambda) = \text{diag}(R(\lambda_1), R(\lambda_2), \dots R(\lambda_n))$.

2.3 Katz Centrality for Discrete Time Evolving Networks

Here we generalize the notion of vertex-to-vertex Katz centrality that we defined above for static networks, like friendship networks. We consider evolving networks that represent time-stamped, peer-to-peer communications. In dealing with such networks, we have to consider walks through them with successive edges being at the same or later time steps.

Sending messages across such networks is obviously asymmetric, even when each individual communication is symmetric. If I talk to you today and you talk to Ramona tomorrow, then I could have sent a message via you to Ramona, but she could not have sent a message to me. So unlike static networks, the passage of time produces asymmetries, and good senders/transmitters are not necessarily the same as good listeners/receivers.

Suppose that we wish to find out who is a good transmitter of information through an evolving network, and who is a good receiver, who has access to all the information being passed around an evolving network. Perhaps we wish to make some intervention. For a static network these properties are the same, and vertices could be rank ordered by the row/column sums of the symmetric centrality matrix. Yet this will not be the case here.

Consider an evolving network (represented by its evolving adjacency matrix), say $\{A_k\}$, for $k = 1, 2, \ldots, K$ on an enumerated set of vertices, $\{1, 2, \ldots, n\}$. The sequence lies in $\mathcal{A}$. We will say the evolving network is equal to A_k at time step t_k.

A *dynamic walk* of length m from vertex i_1 to vertex i_{m+1} is a sequence of edges $(i_1, i_2), (i_2, i_3), \ldots, (i_m, i_{m+1})$ and a non-decreasing sequence of times

$$t_{r_1} \leq t_{r_2} \leq \ldots \leq t_{r_m},$$

such that the edge (i_j, i_{j+1}) is traversed at time t_{r_j}. This is equivalent to saying that

$$(A_{r_j})_{i_j i_{j+1}} = 1.$$

Hence the dynamical walk is made up of m successive edges, each drawn from the evolving network at some suitable non-decreasing set of time steps. In this definition we can use successive edges at the same time or at later times.

The time-dependent context offers a rich variety of alternatives. In some circumstances it may be appropriate to allow at most one edge to be used per time point, so the constraint on the time sequence would involve strict inequality $t_{r_1} < t_{r_2} < \ldots < t_{r_m}$. For example, at a formal evening ball, each time the band plays a dance, you may partner-up with an acquaintance or sit it out. (No cutting in is allowed!) How could you spread a rumour from partner to partner?

Alternatively, it may be appropriate to force exactly one edge to be used per time point, so that we have $r_{j+1} = j + 1$. Analogous definitions of dynamic paths and dynamic trails could be made.

Here we will keep to our original definition $t_{r_1} \leq t_{r_2} \leq \ldots \leq t_{r_m}$, so that any number of the edges could be selected for a single time step, but the ordering of the edges cannot go back in time.

The dynamic walk concept proceeds via the observation that the matrix product

$$A_{r_1} A_{r_2} \ldots A_{r_m}$$

has (i, j)th element that counts the number of dynamic walks of length m from vertex i to vertex j, where the jth step of the walk takes place at time t_{r_j}. This follows from the theorem proved in Chapter 1.

Consider all of the possible dynamical walks from vertex i to vertex j that take exactly $m_k \geq 0$ edges at each time step $k = 1, \ldots, K$. Such a walk is of total length $m = m_1 + \cdots + m_K$, where the walk traverses exactly $m_k \geq 0$ edges from A_k. The number of such walks is given by the (i, j)th term of the matrix product

$$A_1^{m_1} A_2^{m_2} \ldots A_K^{m_K}.$$

Just as in the static network case, we want to count up all such walks that are possible (by varying the m_k), though we will discount each walk according to α^m, where m is the total length.

It follows that the matrix product

$$(I + \alpha A_1 + \alpha^2 A_1^2 + \cdots)(I + \alpha A_2 + \alpha^2 A_2^2 + \cdots) \cdots (I + \alpha A_K + \alpha^2 A_K^2 + \cdots)$$

contains all such terms. So its (i, j)th term counts all possible walks of all possible combinations of edges taken from the successive time steps, suitably discounted for total length.

The matrix product

$$Q = (I - \alpha A_1)^{-1}(I - \alpha A_2)^{-1} \cdots (I - \alpha A_K)^{-1} \tag{2.1}$$

is called the *centrality matrix* for the evolving network $\{A_k \,|\, k = 1, 2, \ldots, K\}$. It does the job of collecting together all dynamic walks: its (i, j)th element gives the overall sum of dynamic walks from i to j, each discounted by α^m, where m is the length.

We can write this as the matrix product (of resolvents):

$$\mathcal{Q} = \prod_{k=1}^{K} (I - \alpha A_k)^{-1},$$

providing we expand out the product of successive terms to the right. In order that $\mathcal{Q}$ is well defined, the infinite sums (the inverses deployed) must exist. This requires that $\alpha < 1/\rho(A_k)$ for all $k = 1, 2, \ldots, K$. Hence α is bounded above by the inverse of the maximum spectral radius of the adjacency matrices.

Recall that $\mathcal{Q}_{ij}$ provides a measure of how information is passed from vertex i to vertex j. Let $\mathbf{s} = (1, 1, \ldots, 1)^T$ as before, then we can measure the (relative) strength of the vertices as transmitters of information to, and receivers of information from, all other vertices by

$$\mathbf{b} = \mathcal{Q}.\mathbf{s}, \quad \mathbf{r} = \mathcal{Q}.\mathbf{s}$$

which are vectors of row sums and column sums. These are not equal for an evolving network as they are for a static network. So our chosen vertex for any particular *intervention* depends on what we want to achieve: are we planting rumours, propaganda, or marketing at a strong transmitter? or are we listening to traffic at a good receiver?

Notice that if we only wish to calculate $\mathbf{b}$, and/or $\mathbf{r}$, we may do so without calculating $\mathcal{Q}$. This is very important when n is large.

For $\mathbf{b}$ we calculate backwards in time. We iterate:

$$\mathbf{b}_0 = \mathbf{s}, \quad \mathbf{b}_k = (I - \alpha A_{K-k+1})^{-1} \mathbf{b}_{k-1}, \ k = 1, \ldots, K.$$

Then $\mathbf{b}_K = \mathbf{b}$. Each iteration requires a single sparse solve, since the As will be sparse. Thus we can hope that calculation of $\mathbf{b}$ may be efficient and $O(nK)$, and avoid the calculation of, or even the need to store of, $\mathcal{Q}$ itself.

For $\mathbf{r}$ we take the transpose of $\mathcal{Q}$ and may calculate forwards in time, iterating:

$$\mathbf{r}_0 = \mathbf{s}, \quad \mathbf{b}_k = (I - \alpha A_k)^{-1} \mathbf{r}_{k-1}, \ k = 1, \ldots, K.$$

Then $\mathbf{r}_K = \mathbf{r}$. Again, each iteration requires a single sparse solve, and we avoid Q itself.

This definition is a natural dynamical generalization of the Katz centrality matrix we defined for a static network. Further, even though each resolvent (each term in the product) is symmetric, the overall product of resolvents will generally be asymmetric, reflecting the inherent non-commutativity arising from the arrow of time.

Exercise 2.4

Let A_1, A_1, A_3, A_4 be given by:

$$\begin{pmatrix} 0 & 0 & 1 \\ 0 & 0 & 1 \\ 1 & 1 & 0 \end{pmatrix}, \begin{pmatrix} 0 & 1 & 1 \\ 1 & 0 & 0 \\ 1 & 0 & 0 \end{pmatrix}, \begin{pmatrix} 0 & 0 & 0 \\ 0 & 0 & 1 \\ 0 & 1 & 0 \end{pmatrix}, \begin{pmatrix} 0 & 1 & 0 \\ 1 & 0 & 0 \\ 0 & 0 & 0 \end{pmatrix}.$$

Let $\alpha = 1/4$: why do we know that $\rho(A_k) < 2$? Find Q. Hence show that vertex 3 has the highest transmit centrality (given by the elements of $\mathbf{b}$) and that vertex 2 has the highest receive centrality (given by the elements of $\mathbf{r}$).

Exercise 2.5

Let us consider how the asymmetry of Q arises, and hence the differences between $\mathbf{b}$ and $\mathbf{r}$. Suppose $\{A_k | k = 1, \ldots K\}$ is a given evolving network of undirected pairwise communications (the A_ks are all symmetric).

For any $n \times n$ matrix, M, we define $\mathcal{S}(M) = \frac{1}{2}(M + M^T)$ and $\mathcal{AS}(M) = \frac{1}{2}(M - M^T)$ to be the projections of M onto the space of symmetric matrices and the orthogonal space of anti-symmetric matrices, respectively. It is clear that $M = \mathcal{S}(M) + \mathcal{AS}(M)$, while $\mathcal{S}(\mathcal{AS}(M)) = \mathcal{AS}(\mathcal{S}(M)) = 0$ (and $\mathcal{S}(\mathcal{S}(M)) = \mathcal{S}(M)$ and $\mathcal{AS}(\mathcal{AS}(M)) = \mathcal{AS}(M)$): hence they are orthogonal projections. We say that $\mathcal{S}(M)$ is the symmetric part of M and $\mathcal{AS}(M)$ is the antisymmetric part of M.

The anti-symmetric part of Q governs the differences between the column and row sums of Q. Then $2\mathcal{AS}(Q)\mathbf{s} = \mathbf{b} - \mathbf{r}$.

Write $\mathcal{S}(Q)$ and $2\mathcal{AS}(Q)$ as asymptotic expansions in α up to $O(\alpha)$ and $O(\alpha^2)$, respectively.

2.4 Corporate Email Communications as an Evolving Network

One widely studied time-dependent interaction data set in the public domain lists the email activities of 151 Enron employees (see <http://www.cs.cmu.edu/enron/>).

We first summarize all of the employee-employee email interactions over a period of 1,138 consecutive days. This leads to a sequence of 1,138 symmetric adjacency matrices of dimension 151 × 151.

Figure 2.1 shows the total number of edges per day. In Figure 2.2, just for the purposes of illustrating the dynamic over time, we lump the first 1,080 days into twelve periods of thirty-day 'months' and display the corresponding monthly adjacency matrices.

Returning to the 1,138 single-day adjacency matrices, we know that many of them are empty. In figure 2.3 we contrast the centrality vectors, **b** and **r**, with the total vertex degrees (the sum of degrees over all days). Clearly some very influential people possess only lowish degrees.

Interestingly we know the actual identities of the individuals in this data set as it is a matter of public record (due to the legal proceedings). Obviously, those with high degrees have a likelihood of having high centralities. But it depends how they are distributed within the network, and when (earlier or later in the sequences).

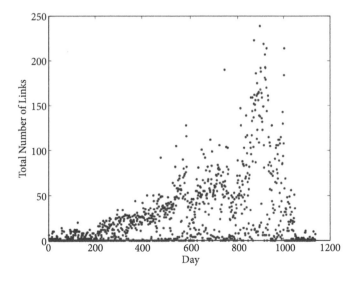

Fig 2.1 Daily edge counts.

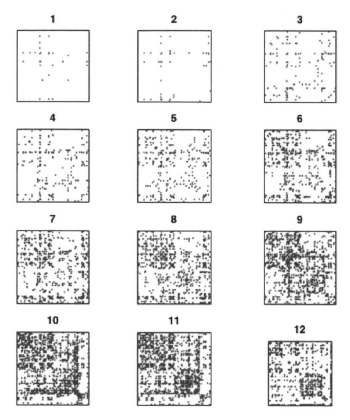

Fig 2.2 Monthly adjacency matrices.

2.5 A Mobile Telephone Call Network

We consider the 52-week Reality Mining data set [37] as an evolving network with $n \times n$ adjacency matrices $\{A_k\}_{k=1}^K$ where $K = 52$ and $n = 106$. Specifically, we used the data on voice calls made between pairs of subjects (106 members of MIT faculty) to derive the weekly adjacency matrices. Figure 2.4 depicts the first 48 weeks of the sequence below. It is very sparse at both the beginning and the end of the sequence, which represents an academic year starting in July.

So for $\mathcal{Q}$ to be well-defined, $1/\alpha$ must be larger than the spectral radius (the Perron–Frobenius eigenvalue) of each of the A_k (larger than 17.456 in this example, which occurs when the evolving network is relatively less sparse). If $\alpha \to 0$, then $\mathcal{Q} = I + \alpha(A_1 + \cdots + A_{52}) + O(\alpha^2)$.

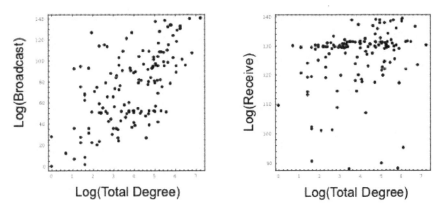

Fig 2.3 Vertex scatterplots of broadcast/source-centrality, **b**, and receive/sink-centrality, **r**.

In Figure 2.5 we show $\mathcal{Q}$ for various values of α.

Figure 2.6 we plots source-centralities held in **b** versus sink-centralities held in **r**. It shows that there are clearly some vertices that are much better as broadcasters/sources than receivers/sinks and vice versa.

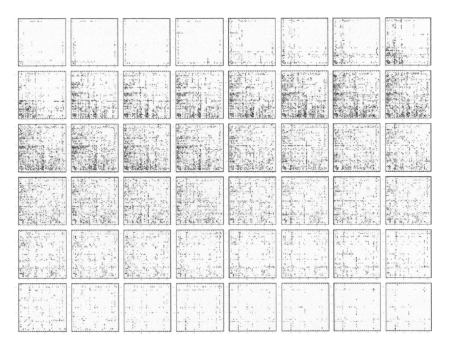

Fig 2.4 Weekly peer-to-peer voice call contacts.

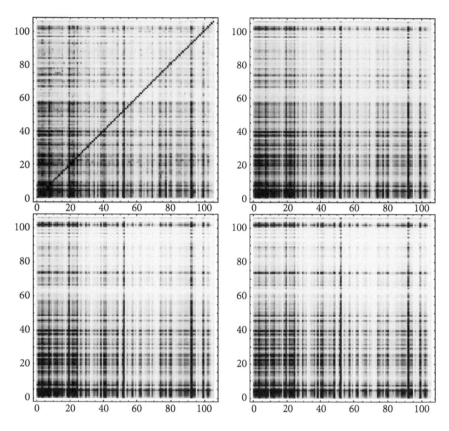

Fig 2.5 $\mathcal{Q}$ for $\alpha = 0.01, 0.02, 0.03, 0.04$.

2.6 Fragility of Evolving Networks

We shall assume that n, the number of vertices, is large. We wish to examine how robust or fragile that network is to insults (knockouts).

Now

$$\mathbf{s}^T \mathbf{r} = \mathbf{s}^T \mathbf{b} = \sum_{i=1, j=1}^{n.n} \mathcal{Q}_{ij}.$$

So we may use this non-negative quantity, denoted by $H(\mathcal{Q})$, as a norm of the total capacity for information to flow from any vertex to any other vertex by any and every possible walk (suitably discounted). This fact was used recently [63]

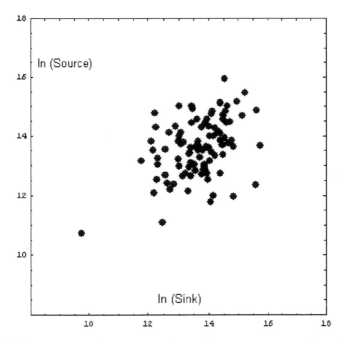

Fig 2.6 Source-centrality, **b**, versus sink centrality, **r**: $\alpha = 0.03$, natural log scale.

to characterize networks by their resilience to random deletions, and introduce a notion of network *fragility*.

Suppose we pick a vertex at random and delete all the edges from every time step that connected to it, to form a new network. So the vertex becomes completely isolated. Imagine calculating a new centrality matrix, Q', for this new network. Clearly it contains an empty row and empty column (except for a one on the main diagonal), whereas Q most likely had non-negative entries there. Moreover other elements of Q' will be less than the corresponding elements of Q since we can no longer count walks that go via the newly-deleted vertex.

Now consider a generated sequence of centrality matrices $\{Q_j | j = 0, \ldots, 1M\}$ where $Q_0 = Q$, that for the original evolving network, and then successive elements, are generated by randomly selecting a voxel that is 'live' within the previous network, and deleting it. At each iteration we may recalculate the centrality. Here, α remains fixed throughout. The corresponding sequence of norms $\{H(Q_j)\}$ is monotonically decreasing. We shall only knock-out $M \ll n$ voxels, which is ideally less than 0.1-1.0% of n, so that there is a low probability of deleting voxels that are highly connected together. We desire that the degradation remains firmly within the linear regime, with small numbers of independent knock-outs. We shall observe the step-by-step degradation of the evolving network, as measured by the monotonic reductions in $\{H(\mathcal{Q}_j)\}$. For some values of j we will annihilate a vertex that plays little role in many walks with the result that $H(\mathcal{Q}_{j-1}) - H(\mathcal{Q}_j)$ is relatively small. For other values we may annihilate a vertex with a large centrality and hence $H(\mathcal{Q}_{j-1}) - H(\mathcal{Q}_j)$ is relatively large.

Suppose the evolving primary network is very lattice-like or homogenous. Then there is a large amount of redundancy in such a network, and since the lattice is relatively homogeneous, almost every random knock-out will produce a similar reduction in functionalty, and the overall progress will be close to linear. For example, imagine a static network on a grid like the roads of Manhattan. If we knock out almost any intersection, the traffic can drive two further blocks around it and little functionally is lost. On the other hand, suppose the primary networks is very tree-like, with few cycles of any length. Then, when some vertices with high centrality are knocked-out, we would expect to see a large reduction in functionality. Think of the UK railway network, for example; if we knock-out Birmingham New Street the network loses a large amount of functionality, yet if we knock-out Henley-on-Thames virtually nobody will notice.

The random degradation process is also suggestive of early-stage decline or damage of an ageing network. This analogy is particularly useful in considering human brains of course, where early-onset cognitive decline is a major issue of interest. In fact it is clear that we ought to see a range of different experiences of cognitive degradation displayed within ageing populations. Some older people lose cognitive functionality in large but occasional steps (presumably having critical irreplaceable, catastrophic losses), hence being somewhat *fragile*. Yet some people's loss is long-term, and relatively smooth and slow (presumably exploiting some network redundancies, and hence displaying a functional robustness to ageing). Thus the proposed approach is a destructive test for the network, as well as providing an experimental analogue to random degradation through ageing. It may form a basis for a future clinical analysis of fMRI scans.

The nature of the degradation arising from a random sequence of knock-outs of these networks may be characterized by two performance measures: the size (in absolute terms) and the distributed nature (variability) of the sequential reductions in functionality, as measured by the $\{H(\mathcal{Q}_j)\}$.

Let M be the number of voxels removed. Then we calculate the quantity

$$\left(1 - \frac{H(\mathcal{Q}_M)}{H(\mathcal{Q}_0)}\right),$$

which is the fractional loss of functionality (as a result of M voxel knock-outs). If this is small then the M insults have had little impact on the primary network, which must consequently be relatively large (dense with connecting edges and thus alternative walks). If this is large then the M insults have removed a more significant amount of functionality and the primary network must consequently be relatively small.

Next consider the successive fractional losses,

$$\{(H(\mathcal{Q}_{j-1}) - H(\mathcal{Q}_j))/(H(\mathcal{Q}_0) - H(\mathcal{Q}_M))| j = 1, \ldots M\},$$

and suppose that they are sorted into descending order. Then we may plot the cumulative fraction of total loss against the cumulative fraction of the total knockouts (ordered by descending size of loss). This curve lies within the unit square, connecting $(0,0)$ to $(1,1)$, above the diagonal with a negative second derivative. We shall calculate the area under this ROC-like curve (receiver-operator characteristic). It is equal to one half if, and only if, all of the fractional losses are equal (all knockouts produce the same loss). It is equal to exactly one if, and only if, all of the fractional losses are zero except for one which is unity (a single knockout accounts for all of the loss). Heuristically we may say that if this is area is small, and close to a half, then the evolving primary network appears to be homogeneous and lattice-like, with many redundancies, and is thus robust; if this is larger, closer to one, then the evolving network has less redundancy and robustness and is more tree-like, and hence is relatively fragile.

We can consider this pair of performance measures, plotted as a point in the plane, as a summary of the primary network's structure.

Now if we do this calculation many times we will obtain distinct results owing to the random selection of successive knockouts. Thus, by resampling the knockout sequence many times over, we calculate an estimate for the means for both measures, together with estimates for the corresponding ranges sampled on either side; see Figure 2.7. Hence we might compare collections of distinct evolving networks, each of which is represented via two performance measures, together with their corresponding sampling error ranges (achieved with a given number of samples). It will be clear when the variability across the collection of individual primary networks is significantly larger than the sampling error ranges on the point estimates.

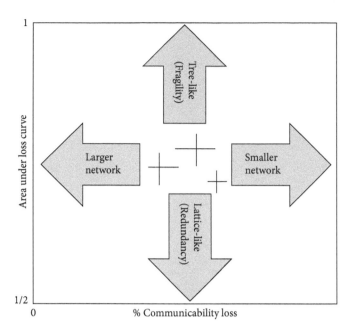

Fig 2.7 Three primary networks plotted as estimates and two-sided ranges, with respect to both performance measures, fragility versus total loss.

In all cases since **r** is relatively cheap, we can calculate all samples of the norms $H(Q_j)$ without holding Q_j itself.

In a paper by Grindrod et al. (2013) [63], these ideas are applied to evolving networks extracted from fMRI scans of around 1,000 human brains (showing the blood oxygen level across the brain, a surrogate for neural activity). Each scan gives rise to an evolving network: fMRI scans are in fact a set of sequential scans over time. Typically each network has hundreds of thousands of vertices representing small voxels (small volumes of brain matter) for which the temporal activity is known. For each of 10^3 brains an evolving network was obtained, and the calculation of each $H(Q_j)$ with $j = 1, \ldots, M = 1,000$ independent knockouts, for each of the networks typically containing 10^5 voxels. All of these computations were carried out using cloud computing resources, see Grindrod et al. (2013) [63] for details. We show some of the results in Figure 2.8, including the fMRI centres where the brains were scanned. The sampling errors on each of the performance measures plotted are far, far smaller than the differences excited across the cloud of points (even for single fMRI centres). Thus the differences between individuals are large compared to possible sampling errors induced by the random knockouts.

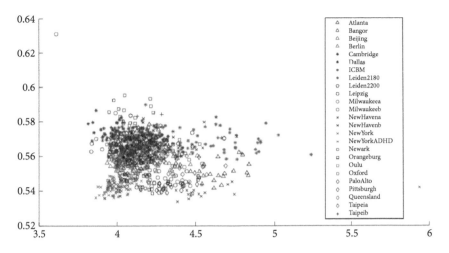

Fig 2.8 Scatterplot showing the fragility measure versus total degradation (1,000 knockouts) for evolving networks defined for 967 brains, indexed by investigating laboratory [63].

2.7 Katz Centrality for Discrete Time Evolving Networks with Age Discounting

Suppose that we receive more and more data as time evolves. Then it is necessary to update $\mathcal{Q}$ for successive time steps. Suppose for ease that each time step is of duration δt.

Let us rewrite (2.6) so that

$$\mathcal{Q}(T) = (I - \alpha A_1)^{-1}(I - \alpha A_2)^{-1} \ldots (I - \alpha A_K)^{-1} \qquad (2.2)$$

where $T = k\delta t$. Then if new information A_{K+1} corresponding to the time interactions within the time step $(K\delta t, (K + 1)\delta t]$, we have

$$\mathcal{Q}(T + dt) = \mathcal{Q}(T)(I - \alpha A_{K+1})^{-1}.$$

So Q is updated by a single (sparse) matrix solve.

In a paper by Grindrod and Higham (2013) [56], a discount was introduced as follows. This is important as it means Q will be successively less dependent on older data, by down-weighting those dynamic walks by their total age (the total time since starting). This is achieved by discounting non-trivial paths at each time step: we have $\mathcal{Q} = I + (\mathcal{Q} - 1)$, where I counts the trivial walk (staying stationary up to now), while $\mathcal{Q} - I$ counts walks with at least one edge from some past time step. So we may rewrite (2.2) as

$$\mathcal{Q}(T + dt) = (I + e^{-b\delta t}(\mathcal{Q}(T) - I))(I - \alpha A_{K+1})^{-1}. \tag{2.3}$$

Here $b > 0$ and any path of length m starting out exactly r time steps ago will be discounted with a weighting equal to $\alpha^m e^{-br\delta t}$.

In a test discussed in Grindrod and Higham (2013) [56] it was shown that if we exploit $\mathcal{Q}$ to forecast some element of the future behaviour of the evolving network, then there is an optimal range of values for the discount parameter. If we take $b = 0$ then all of the history appears equally within $\mathcal{Q}$ and may affect the forecast, but if the evolving network is changing rapidly, or is even non-stationary, this will not be a good thing. On the other hand if we take b to be large, then $\mathcal{Q}$ 'lives in the moment' and there is no memory at all: hence forecasts are likely to be extremely sensitive.

The form given in (2.3) is extremely useful in applications and some experimentation should be made with b depending on the desired use of $\mathcal{Q}$.

2.8 Katz Centrality for Evolving Networks Over Continuous Time

Suppose we can observe our evolving network over continuous time rather than discrete time. In effect we have a time dependent adjacency matrix $A(t)$ where the elements switch from zero to one, or vice versa. It is defined for $t \geq 0$ (or on a suitable interval) taking values in $\mathcal{A}$, and we shall assume that it is continuous from the right (so that we can take limits). Here we follow Grindrod and Higham (2014) [57].

Suppose first that we somehow sample from $A(t)$ so as to discretize the continuous time evolving networks and try to calculate $\mathcal{Q}$.

Fix $\delta t > 0$. Then we might define A_k to be the union of all of the connections existing at any time within the half open interval $((k - 1)\delta t, k\delta t]$. We could symbolically write

$$A_k = \cup_{t \in ((k-1)\delta t, k\delta t]} A(t),$$

then as $\delta t \to 0$ while $K \to \infty$, we will have $A_K \to A(K\delta t^+)$ and we recover an instantaneous sample.

Now consider $\mathcal{Q}$ as in the last section, where we write $T = K\delta t$ and we let $T' = T + \delta t$:

$$Q(T') = (I + e^{-b\delta t}(Q(T) - I))(I - \alpha A_{K+1})^{-1}. \tag{2.4}$$

We cannot take the limit as $\delta \to 0$ in this equation as it is. If we reduce δt by dividing by m then we obtain m new terms each in the form $(I - \alpha A_{K'})^{-1})$ as we update from T to T'. If $A(t)$ is constant between T and T' then it is clear that this limit cannot converge. We would repeatedly update by multiplying by new resolvent terms $(I - \alpha A)^{-1}$ for intervals that are arbitrarily small. To get around this we must re-scale (renormalize) our formula so that it makes sense in the limit.

We modify (2.4) to become

$$Q(T') = (I + e^{-b\delta t}(Q(T) - I))(I - \alpha A_{K+1})^{-\delta t}. \tag{2.5}$$

Then as $\delta t \to 0$, we are updating with successive terms that tend to the identity.

Notice that now we are employing the fractional power of a matrix, in this case fractional powers of the resolvent $(I - \alpha A)^{-1}$, which is itself a matrix valued function of a matrix. We have introduced functions of matrices in an earlier section. If we can define the logarithm of a matrix, $\log(I - \alpha A)$, then we have $\exp(r \log(I - \alpha A)) = \exp(\log(I - \alpha A)^r)$ for both positive and negative values of $r \leq 1$. For our purposes $A(t)$ is always normal and non-negative with spectral radius $\rho(A) < 1/\alpha$. So the spectrum of $I - \alpha A$, which is also normal, is real and lies to the right of zero along the positive axis in the complex plane. So all powers $r \leq 1$, both positive and negative, of $(I - \alpha A(t))$ exist, as does its matrix logarithm.

From (2.5) we have

$$Q(T') = (I + e^{-b\delta t}(Q(T) - I)) \exp(-\delta t \log(I - \alpha A_{K+1})). \tag{2.6}$$

Now we can allow $\delta t \to 0$ in order to obtain an ordinary differential equation (ODE) for a continuously defined centrality matrix:

$$\frac{dQ(T)}{dt} = -Q(T) \log(I - \alpha A(T)) + b(I - Q(T)). \tag{2.7}$$

In order to focus on vertex properties, we can take row and column sums in the matrix $Q(T)$ to define the dynamic broadcast and the dynamic receive centrality vectors just as before:

$$\mathbf{b}(t) = Q\mathbf{s}, \quad \mathbf{r}(t) = Q^T\mathbf{s},$$

where, as usual, $\mathbf{s} = (1, 1, \ldots, 1)^T \in \mathbb{R}^n$.

It is interesting to note from (2.7) that the receive centrality satisfies its own vector-valued ODE

$$\mathbf{r}'(t) = b(\mathbf{s}(t) - \mathbf{r}(t)) - (\log(I - \alpha A(t)))^T \mathbf{r}(t),$$

with $\mathbf{r}(0) = \mathbf{s}$, which is a factor of n smaller in dimension than (2.6) and hence cheaper to calculate. By contrast, it is not possible to disentangle the broadcast centrality, $\mathbf{b}$, in this way. Intuitively, this difference arises because the node-based receive vector $\mathbf{r}(t)$ keeps track of the overall level of information flowing into each vertex and this can be propagated forward in time as new links become available. However, the broadcast vector $\mathbf{b}(t)$ keeps track of information that has flowed out of each vertex: it does not record where the information currently resides and hence we cannot update it based on $\mathbf{b}(t)$ alone. So, with this methodology, real-time updating of the receive centrality is fundamentally simpler than real-time updating of the broadcast centrality.

We also note that since the sum of the row sums is equivalent to the sum of the column sums, and equal to the overall dynamic broadcast centrality of the network, we can monitor the current broadcast ability of the whole network by solving an ODE system that is a factor of n smaller than the system required for nodal broadcast information.

This approach dramatically enhances the existing snapshot-based paradigm for network centrality in terms of both data-driven simulation and theoretical analysis. Our framework fits naturally into the context of online or digital recording of human interactions. The ODE setting conveniently avoids the need to discretize the network data into pre-defined time windows, an approach that can introduce inaccuracies and computational inefficiencies. By defining a continuous time dynamical system, we can solve with an off-the-shelf adaptive numerical ODE solver, so that time discretization is performed 'under the hood', in a manner that automatically handles considerations of accuracy and efficiency. In this way, we may deal adaptively with dramatic changes in network behaviour.

A Personal View

On Digital Media and Marketing

It seems the most natural thing to have networks with edges that appear and disappear over time. They represent pairwise, time-stamped interactions, especially useful for peer-to-peer communications on digital social platforms requiring media analytics. In this chapter we have focussed upon symmetric interactions, such as conversations or voice calls. Yet these ideas go over seamlessly to directed interactions: emails, sms, and so forth. In all applications we have centrality matrices, Q, and broadcast and receive centralities that are usually time-dependent. However, one huge issue is that we often know when messages are sent yet we do not know when they were received. We have to make an assumption. Nevertheless, even under such assumptions (that the receiver is well motivated to open emails or texts as soon as humanly possible), centrally measures are extremely useful. We shall see this in Chapter 3.

Centrality ideas were not the earliest motivating concern when I started to think about evolving networks in 2008. At that time there were many research papers on static networks and their nature, but few on evolving networks. In fact most *dynamics* concerned things like sequential (often preferential) attachment (describing how networks may grow, and what that means for their scale-free structure); or else considering dynamics such as random walks on networks (while the underlying network remaining fixed). At first I became preoccupied with modelling, proposing stochastic models that could produce evolving networks, and subsequently with inverse problems, calibrating such models to data. Then, together with Des Higham, Ernesto Estrada and my graduate student Mark Parsons, we hit upon the ordered matrix product of resolvents and published on the generalization of Katz centrality for discrete time-stepping systems.

We also published a more accessible article on the front page of SIAM News which had an immediate impact. A few companies emailed me out of the blue. One such, Bloom Agency, based in Leeds in the north of England, said simply, 'we are a digital media company, we have read your research and we would like to put it to work within our business'. I asked them to come and visit me, and figuring that Leeds was a four hour car journey away, suggested that they come at very short notice so I could test their true interest. Three men from Bloom arrived as invited. It was a hot day so we signed a non-disclosure agreement and then went for some refreshment by the Thames. Over the next few months, Des and I-with members of my group-gave Bloom some help to build their own operational versions of our analytics. In return they shared problems and data back with us. And most essentially they agreed to be transparent and furnish us with a frank account of how our methods were being exploited by them and what their commercial value was, in terms of numbers of jobs created for mathematicians and the consequent business developed in terms of clients and revenues. This is a very good arrangement for both sides, as academic researchers increasingly must tell such 'tales of impact', yet they rarely have any hard evidence. In addition, no end-user commercial entity can rely on prototype research from academic researchers. The ideas and methods need first to be translated into operational modes, so

continued

On Digital Media and Marketing *Continued*

that they need to buy services from a company such as Bloom. Meanwhile our researchers were freed up to make more radical progress. Things went so well over the next few years that we made a short film (see http://tinyurl.com/oky9cll). Since that time, my relationship with Bloom has developed further, and Des and I have worked with the Bloom analytics team on a number of papers and conference papers, and we have gained a closer relationship with that company's management. They have built upon our original research concepts with a number of proprietary methods and measurements for both scale and efficiencies, and have developed intuitive and real-time services.

The revenues generated by the digital marketing industry in the UK exceeded those for TV advertising and marketing a few years ago, and certainly before 2011. Ten to fifteen years ago digital marketing (online advertisement and banners) was just a high-tech cottage industry. At that time only thirty five per cent of homes were connected to the internet in the UK. Now, with tablets and smart phones, we are all connected to everything, all the time, everywhere. So today all advertising and marketing agencies of any size have their own growing digital arms. Once they moved beyond click-throughs, search-term optimization, and funnelling people to clients' websites, they needed to get very serious about almost all of the topics here in this book. As individuals spend more and more of their waking lives connected to each other, influencing and commenting, this will become ever more urgent. Understanding, modelling, analysing, and forecasting behaviour on dynamical social networks is one of the central pillars of this digital industry.

CHAPTER 3

Structure and responsiveness

I n this chapter we shall deal with a range of issues regarding modelling processes that govern peer-to-peer behaviour, and the structure that arises from it. What can such models tell us about data, or about life in general? Sometimes they can tell us why we have observed certain phenomena, or about other phenomena that we might have observed.

We shall also consider the validity of the ideas presented in the Chapter 2, as well as some new and broader questions about what is going on within social media, and how analytics can help us to make wise decisions.

Yogi Berra said 'In theory there is no difference between theory and practice. In practice there is.' A good knowledge of models is really essential when we are dealing with big data. The best analysts have an expectation as to what will happen within any application. Often the task of getting any output at all is immense, so some analysts are pleased to just get a result without error flags. But our modelling expectations can allow us to sense-check such results, and thus locate errors or bugs in codes, or even misunderstandings. From a mathematical perspective we solve the problem at hand on many levels, from the conceptual (in theory and with models) to the practical (the implementation over data).

Mathematical Underpinnings of Analytics. First Edition. Peter Grindrod.
© Peter Grindrod 2015. Published in 2015 by Oxford University Press.

3.1 Twitter and Social Media Analytics

The rise of micro-blogging sites, status updates, and peer-to-peer messaging on social media platforms means that all users are both publishers and consumers of information, ideas, rumours, and reactions.

There are a number of applications for customer-facing businesses that reflect this two-way nature of the social media (digital marketing) channel. Fundamentally it differs from all other advertising channels that are broadcast (hopefully reaching the intended target audiences), with very little coming back. Within social media the information and opinions flow both ways and this makes it compelling to both the participants and those companies who target those consumers.

The rise of digital media marketing companies has indeed been phenomenal in the past few years. By 2011, the annual investments in digital marketing in the UK topped that of television advertising. Yet those traditional channels are very well-defined and the returns on investments (ROIs) are understood and regulated. When you advertise during the advert break in a television soap opera, you know what you are buying (some percentage of your target audience will see the advert a number of times), and it is monitored with fixed panels. If you do not get the audience on the panel that you bought, the television company must show the advert again. But what exactly do you get from digital marketing campaigns?

This raises a number of important questions about participation, peer-to-peer influence, and measurable changes in awareness and possibly in behaviour. Here we will highlight a number of these issues, relying on various ideas from analytics discussed earlier, and will address some problems from a mathematical perspective.

3.2 Modelling Insights

Before we consider some data-driven problems, we shall begin with some important insights to be gleaned from attempts to model peer-to-peer communication, social influence, and the formation of behavioural norms. In particular, we shall see that dynamical and stability considerations apply, which may mean that although we might observe a particular phenomenon (such as growth or prevalence of social structure) there may be other (stable) possibilities which we could have observed, or that we may experience on similar occasions.

We shall also consider how stability considerations may mean not only that social systems do not settle down around some single long-term equilibrium (with stochastic fluctuations), but they may constantly evolve, on a global scale, and exhibit chaos. The consequences are that although we might forecast the qualitative behaviour of populations, we should not expect to forecast the behaviour of individuals within such populations.

Dynamics on Networks: One-Way Coupling

Suppose we consider the transmission of an idea, a rumour, or social attitude that once gained will never be lost. Starting out at a point (or points) in a social network, we may consider the dynamic of propagation on that network. Here we will not consider an underlying evolving network, though similar ideas apply: instead we shall assume that the network is static and undirected, and that the propagation of the quality in question does not affect that network. The simplest way to model this dynamic is to consider a random walk over the network.

Let A denote the $n \times n$ adjacency matrix and recall that d_i is the degree of vertex i, and D is the diagonal matrix containing the vertex degrees. Let $\Delta = D - A$ denote the Laplacian (as in Chapter 1). Consider a random walk on the network that starts out at some vertex selected by a given probability distribution $\mathbf{z}_0$ in the simplex

$$\mathcal{W} = \{\mathbf{w} \in \mathbb{R}^n | \mathbf{w} \geq 0,\ \mathbf{w}^T \mathbf{s} = 1\},$$

the set of unit-mass distributions over n vertices (here $\mathbf{s} = (1, 1, \ldots, 1)^T$, as usual).

After time $t > 0$, the position of the walk may be represented by $\mathbf{z}(t) \in \mathcal{W}$. The dynamic of $\mathbf{z}(t)$ is given by

$$\frac{d\mathbf{z}(t)}{dt} = -\kappa \Delta \mathbf{z},$$

where $\kappa > 0$ is a rate constant, assumed known. This system is derived as follows. Suppose that the random walk is at vertex i at time t, and that j is one of the d_i vertices that are adjacent to vertex i. Then we assume that the probability that the walk propagates from vertex i to vertex j in a small time δt is given by $\kappa \delta t$. So the probability that it remains at vertex i is $1 - d_i \kappa \delta t$. Putting all of this together, for all vertices, and using the fact that the walk's position at time t is distributed, we get

$$\mathbf{z}(t + \delta t) = \kappa \delta t A \mathbf{z}(t) + (I - \kappa \delta t D) \mathbf{z}(t).$$

Now we divide by δt and allow $\delta t \to 0$ to get the differential equation. We should think of the rate constant κ as a 'willingness to transmit' to a friend. It will be topic-dependent (there are some ideas and topics that are harder to propagate), and it has the dimension of 'per unit time'.

We know how to solve this initial value problem with the matrix exponential function. We have:

$$\mathbf{z}(t) = \exp(-\kappa \Delta t)\mathbf{z}_0.$$

This is merely the *diffusion* process on a discrete network. The eigenvalues $0 = \lambda_1 \leq \lambda2 \leq \ldots \leq \lambda_n$ of the Laplacian are positive and real (see Chapter 1). If the network is connected then λ_2, the Fiedler eigenvector, is positive and controls the long-term behaviour of the walk. We will have

$$\mathbf{z}(t) = \frac{1}{n}\mathbf{s} + \text{constant } e^{-\kappa \lambda_2 t}\mathbf{v}_2$$

for large t. Here $\mathbf{v}_2$ is the Fiedler eigenvector, orthogonal to $\mathbf{s}$. This is intuitive because the Fiedler eigenvalue reflects the almost-discontinuities (the bottle necks) of the global network, while it is clear that $(1/n)\mathbf{s}$ is the equilibrium distribution in $\mathcal{W}$.

Since this process is linear, the distribution $\mathbf{z}(t) \in \mathcal{W}$ may represent an ensemble of all possible random walks on the network starting out by being distributed according to $\mathbf{z}_0$.

Exercise 3.1

In the above the total probability of taking a step from vertex i in time δt is approximated by $d_\kappa \delta t$. How does this change if the total probability of a step from vertex i in time δt is $\kappa \delta t$, with the next vertex j being chosen at random from the d_i possibilities (the neighbours of vertex i)?

Evolving Networks Generated by Non-linear Models

In Chapter 2 we considered data in the form of evolving networks. Here, we discuss some models that might produce such data.

Consider a population of n individuals connected through a dynamically evolving undirected network representing pairwise voice calls or online chats. Let $A(t)$ denote the $n \times n$ binary adjacency matrix for this network at time t, having a zero diagonal. In the notation of Chapters 1 and 2 we have $A : \mathbb{R} \to \mathcal{A}$.

Our simplest model for $A(t)$ assumes that it is known up until the present moment and then at future times. $A(t)$ is a stochastic object, defined by a conditional probability distribution over $\mathcal{A}$ (the set of all possible adjacency matrices). Each edge within this network will be assumed to evolve independently over time, though it is conditionally dependent upon the current network, so multiple edges may be conditional upon some common parts of the current structure. Rather than model a full probability distribution over $\mathcal{A}$ for future network evolution, conditional on its current structure, say $P_{[\delta t]}(A(t + \delta t)|A(t), X)$, it is enough to specify its expected value $\langle A(t + \delta t)|A(t), X\rangle$, a matrix containing all edge probabilities, from which edges may be generated independently. We used this fact in Chapter 1.

Now we specify our model for the stochastic network evolution via

$$\langle A(t + \delta t)|A(t), X\rangle = A(t) + \delta t \mathcal{F}(A(t)), \tag{3.1}$$

valid as $\delta t \to 0$. Here the real matrix valued function $\mathcal{F}$ is assumed to take values in $\mathcal{S}$, and thus it is symmetric, it has a zero diagonal, and all elements within the right-hand side will be in $[0,1]$. By convention we use the rule in (6.16) to determine the independent edges in the upper triangular part of $\langle A(t+ \delta t)|A(t), X\rangle$, and then impose symmetry.

We shall fix $\mathcal{F}$ more specifically and write

$$\mathcal{F}(A(t)) = -A(t) \circ \omega(A(t)) + (\mathbf{1} - A(t)) \circ \alpha(A(t)).$$

Here $\mathbf{1}$ denotes the adjacency matrix for the *clique*, as in Chapter 1, where all $n(n - 1)/2$ edges are present (all elements are ones except for zeros on the diagonal) so that $\mathbf{1} - A(t)$ denotes the adjacency matrix for the graph complement of $A(t)$; $\omega(A(t))$ and $\alpha(A(t))$ are both real, non-negative, symmetric matrix functions containing conditional edge death rates and conditional edge birth rates, respectively; and $\circ$ denotes the Hadamard, or element-wise matrix product, where one simply multiplies the corresponding terms in each matrix together.

In many cases we can usefully consider a discrete time version of the above evolution. Let $\{A_k\}_{k=1}^K$ denote an ordered sequence of adjacency matrices in $\mathcal{A}$ (binary, symmetric with zero diagonals) representing a discrete time evolving network with value A_k at time step t_k. Then we shall assume that edges evolve independently from time step to time step, with each new network conditionally dependent on the previous network. A first order model is given by an iterative process (a process Markov, see Chapter 7):

$$\langle A_{k+1}|A_k, X\rangle = A_k \circ (\mathbf{1} - \tilde{\omega}(A_k)) + (\mathbf{1} - A_k) \circ \tilde{\alpha}(A_k). \tag{3.2}$$

Here $\tilde{\omega}(A_k)$ is a real 'non-negative' symmetric matrix function containing conditional death probabilities, each in $[0,1]$; and $\tilde{\alpha}(A_k)$ is a real, non-negative, symmetric matrix function containing conditional edge birth probabilities, each in $[0,1]$. In the notation of Chapter 1 we have $\tilde{\alpha} : \mathcal{A} \to \mathcal{S}$ and $\tilde{\omega} : \mathcal{A} \to \mathcal{S}$.

As in Chapter 1, the edge independence assumption implies that $P(A_{k+1}|A_k)$ can always be reconstructed from the expected value $\langle A_{k+1}|A_k, X \rangle$.

In the sociology literature, the simplest form of non linearity occurs when people introduce their friends to each other. So, in (6.17), if two non-adjacent people are connected to a common friend at step k, then it is more likely that those two people will be directly connected at step $k + 1$. To model this *triad closure* dynamic [60] we may use $\tilde{\omega}(A_k) = \gamma\mathbf{1}$, so all edges have the same step-to-step death probability, together with

$$\tilde{\alpha}(A_k) = \delta\mathbf{1} + \epsilon A_k^2 \circ \mathbf{1}.$$

Here δ and ϵ are positive and such that $\delta + \epsilon(n-2) < 1$. The off-diagonal elements $(A^2)_{ij}$ count the number of mutual connections that person i and person j have at step k (see Chapter 1). The Hadamard multiplication by $\mathbf{1}$ in $A_k^2 \circ \mathbf{1}$ simply zeros the main diagonal in that term. It is thus clear that both $\tilde{\alpha}(A_k)$ and $\tilde{\omega}(A_k)$ take values in $\mathcal{S}$ as required.

We have

$$\langle A_{k+1}|A_k, X \rangle = A_k \circ (1 - \gamma)\mathbf{1} + (1 - A_k) \circ (\delta\mathbf{1} + \epsilon A_k^2 \circ \mathbf{1}). \qquad (3.3)$$

The resulting dynamical equation is *ergodic*, meaning that over long time, although it may visit all possible states, such instances are weighted and its average behaviour over time can be described by its expected evolution. In this case, the solution is destined to spend most of its time close to states where the density of edges means that there is a balance between edge births and deaths.

A *mean field* approach can be applied by approximating each A_k with its own expectation, which (by symmetry) may be assumed to be of the form $p_k\mathbf{1}$, an Erdős–Rényi random graph, with edge density p_k. We may make such an assumption since no vertices are preferred by the model and the model is invariant under permutations of the vertices (see Exercises 3.4).

In the mean field dynamic one obtains (see Exercise 3.2)

$$p_{k+1} = p_k(1 - \gamma) + (1 - p_k)(\delta + (n - 2)\epsilon p_k^2). \qquad (3.4)$$

Notice that the mean field dynamic is deterministic, since we have averaged out the stochastic fluctuations. The original stochastic dynamic can, and will, deviate (relatively) from this, but may do so with a relatively small probability, and thus only relatively occasionally.

If δ is small and $\omega < \epsilon(n-2)/4$ then this nonlinear iteration has three fixed points: two of them stable, at $\delta/\gamma + O(\delta^2)$ and $1/2 + \sqrt{1/4 - \gamma/\epsilon(n-2)} + O(\delta)$; and one unstable in the middle at $1/2 - \sqrt{1/4 - \gamma/\epsilon(n-2)} + O(\delta)$. Thus the extracted mean field behaviour is bistable.

In practice, an observer of data from such a system would see the edge density of such a network approaching one or other stable mean field equilibria, and jiggling around it for a very long time, without any awareness that another type of orbit or pseudo stable edge density could exist. That is why models are so important, as they imply possibilities or consequences that we have not yet seen.

Direct comparisons of transient orbits from (6.17), incorporating triad closure, with their mean-field approximations in (3.4) are very good over short to medium time scales. Yet though we have captured the nonlinear effects well in (3.4), the stochastic nature of (6.17) must eventually cause orbits to diverge from the deterministic stability seen in (3.4).

Exercise 3.2

Assuming $\langle A_k | X \rangle = p_k \mathbf{1}$ and $\langle A_{k+1} | A_k, X \rangle = A_k \circ (1 - \gamma)\mathbf{1} + (1 - A_k) \circ (\delta \mathbf{1} + \epsilon A_k^2 \circ \mathbf{1})$ then show directly that $\langle A_{k+1} | X \rangle = p_{k+1}\mathbf{1}$, where p_{k+1} is given by $p_{k+1} = p_k(1 - \gamma) + (1 - p_k)(\delta + (n-2)\epsilon p_k^2)$.

Exercise 3.3

The model in (3.3) can represent a new year cohort in a graduate school where few of the n students know any of the others, and where we measure the social graph of student friendships weekly. Assume $\delta \to 0$. If triad closure (introducing friends of friends) never gets going, then we are destined to approach the lower equilibrium where random friendships form at a rate close to δ and existing friendships expire at a rate close to γ. The mean field model (3.4) applies. If there are cider and cheese parties early on in the first term, the system may get shifted to the upper, more sociable, equilibrium. How many friendships will we expect to observe at any subsequent week close to equilibrium? Assuming that A_k is well approximated by the equilibrium density Erdős–Rényi random graph, what is the probability that there will be zero friendships (a social catastrophe) the very next week? If $n = 50$, $\epsilon = 1/200$, and $\gamma = 0.2$, what is this probability?

For any $n \times n$ matrix, B, and any permutation matrix, Q, the matrix QB permutes the rows of B; while the matrix BQ^T similarly permutes the columns. The matrix QBQ^T permutes both rows and columns of B.

(a) If $\mathcal{F} : S \to S$ is such that no vertices are *preferred*, meaning that if we re-enumerate the vertices the evolution will remain the same, then if we permute the vertices we must retain the same dynamic. Show that in that case $\mathcal{F}$ must satisfy

$$\mathcal{F}(A) = Q^T \mathcal{F}(QAQ^T)Q$$

for all permutation matrices Q and all $A \in S$. This condition is not as restrictive as it appears. It is a *symmetry condition*.

(b) Show that power of A and thus any polynomial in A, $\mathcal{F}(A) = r_0 I + r_1 A + r_2 A^2 + \cdots + r_m A^m$, has this property.

(c) Show that the Hadamard product of any two polynomials in A has this property.

(d) Show that

$$A \circ (1 - \gamma)\mathbf{1} + (1 - A) \circ (\delta \mathbf{1} + \epsilon A^2 \circ \mathbf{1}),$$

which we used in (3.3), has this property.

Finally suppose that a population is inhomogeneous, so that some individuals are not exchangeable, and thus not permutable. We might have a mix of men and women, or officers and men, or staff and students, for example. It would be reasonable to suppose that (6.17) should be extended so as to reflect this. The point is that male-male friendships may behave differently from male-female and female-female friendships. Suppose that the n vertices are ordered so that the first n_1 individuals are male and the next $n_2 = n - n_1$ individuals are female. Let us write

$$\mathcal{F}(A_k) = A_k \circ (1 - \tilde{\omega}(A_k)) + (1 - A_k) \circ \tilde{\alpha}(A_k).$$

Then (6.17) becomes

$$\langle A_{k+1} | A_k, X \rangle = \mathcal{F}(A_k).$$

Now $\tilde{\alpha}(A_k)$ and $\tilde{\omega}(A_k)$ require some block diagonal structure so that for example different constants ϵ, δ and γ apply to the three different types of adjacency. There will be three sets of values, one for male-male friendships in the first $n_1 \times n_1$ diagonal block, one for female-female friendships in the last $n_2 \times n_2$ diagonal block, and one for the mixed sex friendships in the first $n_1 \times n_2$ off-diagonal block. For example we might have

$$\tilde{\alpha}(A_k) = \tilde{\delta} + \tilde{\epsilon} \circ A_k^2 \quad \tilde{\omega}(A_k) = \tilde{\gamma},$$

where $\tilde{\epsilon}$, $\tilde{\delta}$ and $\tilde{\gamma}$ are matrices in $\mathcal{S}$ with that block structure, and thus three degrees of freedom. In that case, and more generally, $\mathcal{F}(A_k)$ must be invariant under permutations that permute the subset of men and permute the subset of women separately. This has implications for the mean field approach. Since not all permutations are allowable ($\mathcal{F}$ is only invariant to perturbations that preserve gender), the Erdős–Rényi graph $p_k \mathbf{1}$ is no longer a suitable candidate for the expected value for A_k.

In practice, if the population under observation is segmented in some way, one might search for evidence that this matters. Worse, if we do not know which segment individuals are in, we might have to do some preliminary work in order to infer this by clustering the individuals on the basis of their observable dynamical behaviour.

Fully Coupled Systems

In our next example we shall consider *full coupling* between the dynamics *of* the network and the dynamics *on* the network, and show how this may give rise to behaviour that is qualitatively predictable at the macroscopic level, but not at the individual level. The network here is not static; it is driven in some way by the dynamic propagating over it. It is important to realize that phenomena based on social influence do not necessarily settle down as they are often compelled to do so by certain classes of model in the literature. Rather, they may remain *boiling*.

There is a large literature within psychology that is based on individuals' attitudes and behaviours being in a tensioned equilibrium between (excitatory) activating processes and inhibiting processes. Typically the state of an individual is represented by a set of state variables, some measuring activating elements and some measuring the inhibiting elements.

Activator-inhibitor systems have had an impact within mathematical models where a uniformity equilibrium across a population of individual systems becomes destabilized by the very act of simple passive coupling between them. Such Turing instabilities can sometimes seem counter-intuitive on first sight. *Homophily* is a term that describes how associations are more likely to occur between people who have similar attitudes and views. Here we show how individuals' activator-inhibitor dynamics coupled with homophilic evolving networks produce systems that have pseudo-periodic consensus and fractionation.

Consider a population of n identical individuals, each described by a set of m real attitude state variables that are continuous functions of time t. Let $\mathbf{x}_i(t) \in \mathbb{R}^m$ denote the ith individual's attitudinal state. Let $A(t)$ denote the adjacency matrix for the communication network, as in (6.16). Then consider

$$\dot{\mathbf{x}}_i = \mathbf{f}(\mathbf{x}_i) + D_f \sum_{j=1}^{n} A_{ij} \left(\mathbf{x}_j - \mathbf{x}_i \right) \quad i = 1, \ldots, n. \tag{3.5}$$

Here, $\mathbf{f}$ is a given smooth field over $\mathbb{R}^m$, drawn from a class of activator-inhibitor systems (see Exercise 3.5, for an example), and is such that $\mathbf{f}(\mathbf{x}^*) = 0$, for some, $\mathbf{x}^*$, and the Jacobian there, $\mathbf{df}(\mathbf{x}^*)$, is a stability matrix, that is, all of its eigenvalues have negative real parts. Here, D_f is a real diagonal, non-negative matrix containing the maximal transmission coefficients (diffusion rates) for the corresponding attitudinal variables between adjacent neighbours. Let $\mathbf{X}(t)$ denote the $m \times n$ matrix with ith column given by $\mathbf{x}_i(t)$, and $\mathbf{F}(\mathbf{X})$ be the $m \times n$ matrix with ith column given by $\mathbf{f}(\mathbf{x}_i(t))$. Then (3.5) may be written as

$$\dot{\mathbf{X}} = \mathbf{F}(\mathbf{X}) - D_f\mathbf{X}\Delta. \tag{3.6}$$

Here, as before, $\Delta(t)$ denotes the graph Laplacian for $A(t)$, given by $\Delta(t) = D(t) - A(t)$, where $D(t)$ is the diagonal matrix containing the degrees of the vertices. This system has an equilibrium at $\mathbf{X} = \mathbf{X}^*$ say, where the ith column of $\mathbf{X}^*$ is given by $\mathbf{x}^*$ for all $i = 1, \ldots, n$.

Now consider an evolution equation for $A(t)$, in the form of (6.16), coupled to the vertex states, given in $\mathbf{X}(t)$:

$$\langle A(t+\delta t)|A(t), X \rangle = A(t)+\delta t(-A(t)\circ(1-\Phi(\mathbf{X}(t))\gamma+(1-A(t))\circ\Phi(\mathbf{X}(t))\delta). \tag{3.7}$$

Here, δ and γ are positive constants representing the maximum birth rate and maximum death rate, respectively, and the homophily effects are governed by the *pairwise similarity* matrix, $\Phi(\mathbf{X}(t))$, such that each term, $\Phi(\mathbf{X}(t))_{ij} \in [0, 1]$ is a monotonically decreasing function of a suitable semi-norm, $\|\mathbf{x}_j(t) - \mathbf{x}_i(t)\|$.

We shall assume $\Phi(\mathbf{X}(t))_{ij} \sim 1$ for $\|\mathbf{x}_j(t) - \mathbf{x}_i(t)\| < \epsilon$, and $= 0$ otherwise, for some suitably chosen $\epsilon > 0$.

There are equilibria at $\mathbf{X} = \mathbf{X}^*$ with either $A = 0$ (the empty network) or $A = \mathbf{1}$ (the full clique). To understand their stability, let us assume that δ and $\gamma \to 0$. Then $A(t)$ evolves very slowly, via (3.7), compared to $\mathbf{X}(t)$, which is governed by (3.6). Let $0 = \lambda_1 \le \lambda_2 \le \ldots \le \lambda_n$ be the eigenvalues of Δ. Then we will show that $\mathbf{X}^*$ is asymptotically stable only if all n matrices, $\mathbf{df}(\mathbf{x}^*) - D_f \lambda_i$, are simultaneously stability matrices; and conversely it is unstable in the ith mode of Δ if $\mathbf{df}(\mathbf{x}^*) - D_f \lambda_i$ has an eigenvalue with a positive real part.

To see this, consider the linearization of (3.6) at $\mathbf{X}^*$, and its behaviour in the ith mode ($i = 1, \ldots, n$). We write $\mathbf{X}(t) = \mathbf{X}^* + \mathbf{y}(t)\mathbf{v}_i^T$ where $\mathbf{y}(t) \in \mathbb{R}^m$ is very small, and $\mathbf{v} \in \mathbb{R}^n$ is the ith eigenvector of Δ. Then, retaining only linear terms in the small perturbation, $\mathbf{y}(t)$, we have

$$\dot{\mathbf{y}}\mathbf{v}_i^T = \mathbf{df}(\mathbf{x}^*)\mathbf{y}\mathbf{v}_i^T - D_f \lambda_i \mathbf{y}\mathbf{v}_i^T,$$

so that

$$\dot{\mathbf{y}} = (\mathbf{df}(\mathbf{x}^*) - D_f \lambda_i)\mathbf{y}.$$

Thus the perturbation in the ith mode will tend to zero if, and only if, $\mathbf{df}(\mathbf{x}^*) - D_f \lambda_i$ is a stability matrix.

Now one can see the possible tension between homophily and the attitude dynamics. If the population begins with uniform attitudes, the density of edges will grow (through (3.7)). But as the Laplacian changes, one or more of the matrices $(\mathbf{df}(\mathbf{x}^*) - D_f \lambda_i)$ may lose stability, in which case patterns across the network will begin to emerge, reflecting the structure of the corresponding eigenmode. When such disparities become large, the homophily terms will flip and some edges will expire.

Consider the spectrum of $\mathbf{df}(\mathbf{x}^*) - D_f \lambda$ as a function of λ. If λ is very small then this is dominated by the stability of the uncoupled system, $\mathbf{df}(\mathbf{x}^*)$. If λ is large then this is again a stability matrix, since D is positive definite.

The situation, dependent on some collusion between choices of D_f and $\mathbf{df}(\mathbf{x}^*)$, where there is a *window of instability* for an intermediate range of λ in which at least one of the eigenvalues has positive real part, is known as a Turing instability. It occurs in a very wide class of activator-inhibitor systems. Note that as $A(t) \to \mathbf{1}$, we have $\lambda_i \to n$, for $i > 1$. So if n lies within the window of instability, then we are assured that the systems can never reach a stable, consensual, fully-connected equilibrium. Instead, Turing instabilities can drive the break-up (weakening) of the network into relatively well-connected subnetworks. These in turn may restabilize the equilibrium dynamics (as the eigenvalues leave the window of instability), and then the whole process can begin again as homophily causes any absent edges to reappear.

The key point of learning is that this can happen for a rather wide class of system dynamics, and we should not assume that such systems can settle down close to an equilibrium.

Thus, we expect a pseudo-cyclic emergence and diminution of patterns, representing transient variations in attitudes. In simulations, by projecting the network $A(t)$ onto two dimensions using the Frobenius matrix inner product, one may observe directly the cyclic nature of consensus and division. This has been demonstrated by Ginndord and Pasons (2012) [62].

If the stochastic dynamics in (3.7) are replaced by deterministic dynamics for a weighted adjacency matrix, one obtains a system that exhibits a-periodic wandering and also sensitive dependence [123]. In such cases the orbits are chaotic, and we know that they will oscillate, but we cannot predict whether any specific individuals will become relatively inhibited or relatively activated within future cycles, Figure 3.1. This phenomenon even occurs when $n = 2$.

These models show that when individuals, who are each in a dynamic equilibrium between their activational and inhibitory tendencies, are coupled in

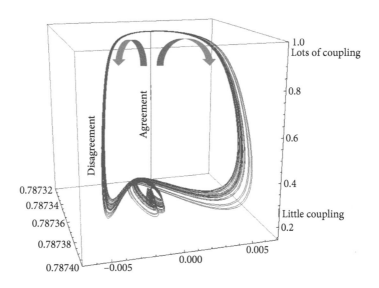

Fig 3.1 Pseudo-cyclic chaotic dynamics for a deterministic model with a weighted adjacency matrix [123]: edge weighting (vertical) versus the differences in vertex-state variables, for a single pair.

a homophilic way, we should expect a relative lack of global social convergence to be the norm. Radical and conservative behaviours can coexist across a population and are in a constant state of flux. Within both deterministic and stochastic versions of the model, the macroscopic situation is predictable, while the journeys for individuals are not.

There are some commentators in socio-economic fields who assert that divergent attitudes, beliefs, and social norms require leaders and are imposed on populations; or else they are driven by partial experiences and events. But here we can see that the transient existence of locally clustered subgroups, holding diverse views, can be an emergent behaviour within fully-coupled systems. This can be the normal state of affairs within societies, even without externalities and forcing terms.

Sociology studies have, in the past, focussed on rather small groups of subjects within experimental conditions. Digital platforms and modern applied mathematics will transform this situation. Computational social science can use vast data sets from very large numbers of users of online platforms (Twitter, Facebook, blogs, group discussions, multiplayer online games) to analyse how norms, opinions, emotions, and collective actions may emerge and spread through local interactions.

Exercise 3.5

The Schnackenberg system is a two-variable, activator inhibitor system that, using Murray's variables [107, 61], is given by

$$\dot{x}_1 = f_1(x_1, x_2) = p - x_1 x_2^2, \quad \dot{x}_2 = f_2(x_1, x_2) = q - x_2 + x_1 x_2^2.$$

Here $p > q > 0$ are constants such that $(p + q)^3 > p - q$. Show that $\mathbf{x}^* = (x_1^*, x_2^*) = (p/(p + q)^2, p + q)$ is a rest point. Linearize the system about that point and show that the trace of the linearization, $\mathbf{df}(\mathbf{x}^*)$, is negative, while its determinant is positive. Hence the equilibrium is asymptotically stable. Now suppose dispersion is added so that the systems is modified:

$$\dot{x}_1 = f_1(x_1, x_2) + d_1(x_1^* - x_1), \quad \dot{x}_2 = f_2(x_1, x_2) + d_2(x_2^* - x_2),$$

for positive constants d_1 and d_2. The linearization of this system about the same rest point, $\mathbf{x}^*$, yields the matrix $\mathbf{df}(\mathbf{x}^*) - D_f$, where $D_f = \text{diag}(d_1, d_2)$. This is our equivalent of the matrix $(\mathbf{df}(\mathbf{x}^*) - D_f \lambda_i)$ for the positive eigenvalues of the Laplacian. The sum of the eigenvalues of this system remains negative, and equal

continued

to the trace of this matrix. Show that it is possible for the determinant to become zero. Hence the equilibrium can lose stability, specifically, provided that $d_2 < (p - q)/(p + q)$ and d_1 is large enough. State variable x_1 is the inhibitor while x_2 is the activator. If dispersion prevents a rising in x_1 above equilibrium from opposing a rising in x_2, then instability follows.

3.3 Validating the Identification of Influencers

How can influencers be identified on large scale, and in real time? The ideas presented in Chapter 2 provide a number of ways that influencers within real social networks can be recognized using the various generalizations of Katz centrality. Here we follow Laflin et al. (2013) [91] and investigate how this stands up in practice.

Consider use of centrality measures to discover influential players in a dynamic Twitter interaction network, with respect to some given topic. Empirical evidence is given in Gleave et al. (2009) [47] for the existence of catalysts in an online community 'responsible for the majority of messages that initiate long threads.' Huffaker [81] identifies online leaders who 'trigger feedback, spark conversations within the community, or even shape the way that other members of a group talk about a topic. Experiments in Mantzaris and Higham (2012) [99] on email and voicemail data found evidence of individuals 'punching above their weight' in terms of having an ability to disseminate or collect information that cannot be predicted from static or aggregate summaries of their activity. A recent business-oriented survey [15] lists network dynamics as a key technical challenge, and Shamma, Kennedy, and Churchill (2011) [115] argue that 'the temporal aspects of centrality are underrepresented.'

Several recent studies have addressed the issue of discovering important or influential players in networks derived from Twitter data. The research in Bakshy et al. (2011) [9] focussed on how a shortened URL is passed through the network (assuming that a person who passes on such a URL has been influenced by the sender), so as to study the structure of cascades. In Cha et al. (2010) [21], a large scale Twitter follower graph is considered and in Kwak et al. (2010) [90] users are ranked by their number of followers, though they found that a retweet measure produced a very different ranking. But none of the influence measures considered in these studies [21, 90] fully respect the time-ordering of Twitter interactions.

The analysis in Laflin et al. (2013) [91] differs from that just described by focussing on subject-specific tweets of interest within a typical marketing application: building the interactions between tweeters as an evolving network; and comparing a range of centrality measures with independent, hand-curated rankings from social media experts exposed to the full tweet contents.

The Twitter feed can be used to construct a topic-based, active, evolving network (removing individuals deemed to be non-active) in the following way.

(a) Over a given period of time, $[t_\alpha, t_\omega]$, first extract all tweets that contain any elements from a set of required phrase(s), defining the topic of specific interest. Each time such a new tweet is recorded, add the sender and all of the sender's followers (who may see the sender's tweets) to the network, and add a time-stamped directed edge from the sender vertex to all follower nodes.

(b) Remove all vertices that have zero aggregate out degrees (they did not send out any relevant tweets and were thus inactive).

(c) Slice the data into M windows of size $\delta t = (t_\omega - t_\alpha)/M$. Denote $t_k = t_\alpha + (k-1)\delta t$, for $k = 1, 2, \ldots, M$, so that the kth time window is $[t_k, t_{k+1})$ and that activity is represented by a binary adjacency matrix A_k describing the union of all edges present within that window.

So after the initial topic-based filtering, no further attention to the content of the tweets is given, and the data extracted in the time-ordered $\{A_k | k = 1, \ldots, M\}$ are purely topological. Note these networks are directed rather than undirected, as the networks were in earlier chapters; though it is clear that the corresponding definitions and methods apply.

In Laflin et al. (2013) [91] this method was used to construct an active evolving network from tweets containing the phrases:

city break, cheap holiday, travel insurance, cheap flight

and two phrases relating to specific travel brands. This simulated a typical client-driven investigation on behalf of a travel company wishing to improve its social media presence. The data collection took place from 17 June 2012 at 14:41 to 18 June 2012 at 12:41 with δ_t equal to 66 minutes, producing $M = 20$ time windows. The total number of tweeters and their followers associated with this small data set was 442,948. Restricting attention to active nodes reduced the network size to $n = 590$.

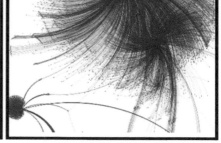

Fig 3.2 Two details from the active network sequence at the end of the first time window: on the left we see a tiny isolated conversation from that window [91].

Some Twitter accounts produced extremely rapid bursts, and one account tweeted 104 times in timeframe ten and a further twenty-three times in timeframe eleven. To give a feel for the data, Figure 3.2 (from the platform developed by Bloom Agency and extracted from Laflin et al. (2013) [91]) visualizes two portions of the network at the end of the first time window.

In Laflin et al. (2013) [91], centrality measures were calculated in different ways: (i) broadcast and receive centralities using the standard Katz centrality for a single static network, representing the union, all interactions (edges) together (taking $\alpha = 0.9/\rho(A)$); (ii) broadcast and receive centralities using the discrete time Katz centrality concept for the evolving network $\{A_k\}$ (taking $\alpha = 0.9/\max_k\{\rho(A_k)\}$); and (iii) in and out degree centrality.

To benchmark these centrality measures, five industry professionals working in social media with day-to-day experience of ranking and targeting accounts based on Twitter data were enlisted. These experts look for influential vertices (accounts) in the network as a means for targeted intervention and engagement. It is not feasible to study the full set of dynamic interaction data by eye across the 590 active vertices. Indeed, this is a key motivation for the use of automated measures and tools. Hence attention was focussed on a list of forty-one accounts that were felt to be highly relevant based on the automated ranking results and further account-level information. The five experts were given access to the full details of the tweets from this list, including the content of their messages, and asked to rank them in order of importance. They had no knowledge of the centrality rankings. In Table 3.1 we show the top ten anonymized account IDs in rank order from one down to ten for each measure. We merged the five different expert rankings of the forty-one vertices, giving equal weight to each, into a single list.

Table 3.1 *The top ten anonymized account IDs in rank order*

Average over five experts	out degree	static broadcast	dynamic broadcast
397	74	74	74
362	34	302	398
398	362	362	362
341	370	358	34
289	358	375	358
345	71	34	302
462	345	341	397
212	398	352	352
71	352	200	373
18	484	409	380

For each of the three centrality categories: degree, static Katz, and dynamic Katz, the broadcast (out) measures in the table performed much better than receive (in) measures. This agrees with our initial intuition that, in the context of targeting influential users whose behaviour affects others in the network, information dispersal is more relevant than information gathering.

We see that dynamic Katz broadcast has a top three that includes two of the experts top three. Out degree and static Katz broadcast have one such correct answer in their top three. Also note that the centrality rankings are closer to each other than to the averaged expert in terms of overlap.

The automation of influencer identification is an important and essential step. Results such as these give us confidence that it can be scaled far beyond the capabilities of subjective domain experts.

3.4 Sampling: When Should You Over-Invest in Social Media Channels?

This section is all about defining useful subgraphs from a given graph.

Taking random sampled subnetworks from a network may be a very useful way of describing and investigating it, but it is an especially useful idea when the samples are selected in a possibly biased manner, by some underlying property. If we compare such subnetworks to those of similar size with vertices that are randomly and independently sampled (a null hypothesis) then we may infer something about that sub-community.

Suppose that we are presented with a static social network from some platform, or that we are able to observe some, or most, of it. The result is a graph containing vertices representing individuals and edges that represent mutual friendship. Let us denote this graph by G. It may be very large indeed for a social network or mobile network operator.

In practice it is not so simple to establish G from static information such as friendships on Facebook or followers on Twitter. The reason is that many people are quite cavalier about who they follow or befriend. Indeed there are many social media accounts that possess far more friends than they could reasonably cope with in any diligent way. Equally friendships may not be reciprocated. A practical solution to this on the Twitter platform is to find connections where individuals mention each other specifically above some frequency. We refer to this as the 'mentions network' as it shows undirected connections between two individuals, where each invests at least some time and effort in informing the other about specific things. On Twitter this involves a direct mention (by @name) within a tweet. Moreover, we only accept connections where we have observed mentions in both directions (above some minimal frequency over some defined period of time). It would be possible for mobile phone operators (MNOs) to construct equivalent networks, formed of pairs people who call or text each other on their network, regularly, above some rate.

In principle we can examine the national mentions network, G, for a specific messaging platform (social media, Twitter, text, etc). Let us suppose that this is so. Now suppose that we are considering the investment of some effort into that platform so as to reach some target group, within some marketing campaign. Is this wise? Would we be better to broadcast on TV, or in the newspapers, or on the back of a bus?

Suppose that we define a subgraph of G by restricting ourselves to edges between vertices that lie in some 'target' subset of the vertex set. For example, we could look at those individuals who had mentioned various chronic diseases, from asthma to leukaemias; or individuals who had mentioned various consumer brands, from Colgate toothpaste to Yorkshire Tea and Jaguar. Call this new subgraph H. It is the observed subset of G of those connections between individuals within the target vertex set, corresponding to the interest group. We might define H in a number of ways, though typically they would have displayed some knowledge or interest, within previous messages that makes them suitable 'targets'.

Now let $\theta(H)$ denote some well-defined, non-negative measurement of the structure observed in H. It might be a connectedness measure, or the edge density, or anything related to the spectrum of the associated (sub)-graph Laplacian. The idea is that θ should be relatively large when H is near clique-like and relatively small when H is edge-empty.

We wish to compare the observed value of $\theta(H)$ with a distribution of other θ values, all achieved under a null hypothesis. In this case, we shall choose the null hypothesis to be that 'membership of the target group is independent of any structure within the observed social network'. So suppose that we randomly generate other subgraphs from G under the null hypothesis, each with the exactly same number of vertices as H. We will select such a new subgraph, say $\tilde{H}$, by selecting vertices chosen independently at random (and thus clearly independent of vertex-to-vertex structure). For each new subgraph, say $\tilde{H}$, we may calculate $\theta(\tilde{H})$. In doing this many, many times over we are sampling the distribution of θ values observable for subgraphs of a given size drawn under the null hypothesis. Finally, we may calculate the probability that $\theta(\tilde{H})$ exceeds or is equal to the value obtained for our actual target set, $\theta(H)$. This last is called the p-value for $\theta(H)$ under the null hypothesis. If this is very small then H contains structure (as measured by θ) far beyond that expected for the random subgraphs. Thus the target group is highly unlikely to be independent of the social structure on this platform, and observed in G. On the other hand, if the probability is larger (say greater than one per cent or five per cent) then there is nothing especially remarkable about the target group in the social context; it is not dissimilar to a random subgraph of the same size drawn from G, and membership is unlikely to be associated with social interactions on this platform.

We may express this the other way around by calculating the values for $\theta(\tilde{H})$ corresponding to fixed p-values in $(0,1)$, as the size of H varies. We obtain a p-value contour in the plot of θ vs size. For distinct p-values, we obtain distinct contours: these contours are all functions of the subgraphs (consisting of independent randomly drawn vertices) of G. We show an example in Figure 3.3 using various chronic diseases and brands to define smallish target sets in each case, over a single mentions network, G, that was defined for 10^5 or so UK Twitter users.

In Figure 3.3 the colours show, likelihood of the structure appearing in a similar sample drawn randomly from a UK Online Social Network (the Twitter Mentions Network). Low p-values (blue contours) indicate brands or chronic diseases that should demand an increased investment in online channels (e.g. Coca-Cola, or various cancers). High p-values (red contours) indicate brands

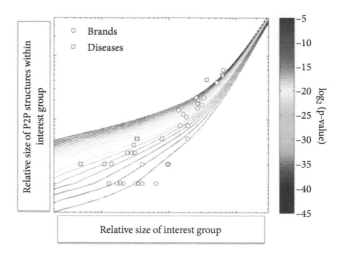

Fig 3.3 Subgraph structure versus subgraph size coloured by *p*-value.

or chronic diseases for which online marketing is no better than broadcasting (e.g. Colgate and asthma), and there is no evidence that the relevant interest groups are structured.

This plot reflects at least two underlying processes. First, that some target groups do have a social structure. People with life-threatening chronic conditions meet each other at clinics, seek advice from each other on the internet, join support groups, and so on. So it is not surprising that these are anomalously clustered within G. On the other hand, asthma is now ubiquitous, sufferers just live with it, and it is not remarkable and does not drive their social interactions. Similarly, the brand of toothpaste we use is not key to forming social relationships.

Other brands such as BMW or Coca-Cola may be of particular interest within certain social strata who are connected to each other for other reasons. So, although not causal, the fact of high connectivity and social structure is still useful in targeting those groups, since there is some possibility of 'buzz' within the target group amplifying the media activity.

3.5 Event-Driven Spikes

The digital footprints left by our online interactions provide a wealth of information and present many challenges for modelling and computation [92]. In particular, microblogging and messaging data offers the prospect of tracking peer-to-peer influence, predicting future behaviour [27] and engaging in targeted intervention [4, 5]. Nowadays, public-facing organizations may interact with their public through 24/7 online global conversations, and may exploit opportunities to leverage current sentiment and interests. Conversations appear like sudden avalanches, triggered by some event and then rolling forwards: huge numbers of individuals can get involved very suddenly.

In this section we follow the analysis by Higham et al. (2014) [77] concerning the situation where a rapid spike of activity can be attributed to a high profile event or news item. For example, in Figure 3.4 we show spikes arising on Twitter during a Bundesliga encounter between Bayern Munich and Borussia Dortmund on 4 May 2013. The vertical axis records the volume of Twitter activity over each one-minute period. Here, a tweet is deemed to take part in the conversation if it contains one or more of a specified set of keywords. These transient bursts of interest represent marketing opportunities for suitably agile companies, as demonstrated by the cookie company Oreo who produced an effective (and award winning) tweet in response to a power failure during Super Bowl XLVII [41].

A number of authors have considered how information is passed in the setting of online social media, focussing on the propagation of *memes* (ideas,

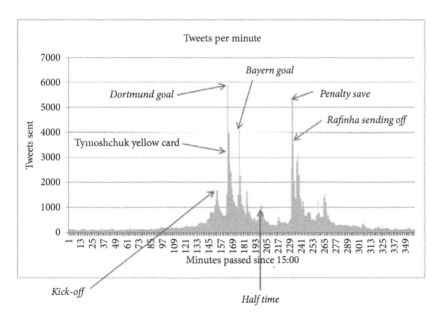

Fig 3.4 The Bundesliga encounter between Bayern Munich and Borussia Dortmund on 4 May 2013, taken from Higham et al. (2014) [77].

behaviours, or styles that spread from person to person) within a closed system. In Cha et al. (2010) [21], the number of followers, retweets and mentions were used to quantify the influence of Twitter users, with the three measures yielding very different results. Similarly, Kwak et al. (2010) [90] ranked users by the number of followers and also by Google's PageRank algorithm. Related work in Lerman, Ghosh and Surachawala (2012) [93] looked at how network structure affects the dynamics of large-scale information flow around news stories. In Centola (2010) [20], the spread of behaviour was examined through artificially constructed, static, social interaction networks, with clustered-lattice structure found to be the most effective in terms of speed and reach. Person-to-person cascades of information spread have been studied by a number of authors [9, 95, 94, 112, 113]; and we recommend Borge-Holthoefer (2013) [16] for an overview of models and applications.

Rather than analysing the development of cascades (the spreading of memes) within an online community based on self-generated content, we shall focus on the event-driven, *full attention span* setting, where the relevant community has been roused by an external development, and some catalytic event takes place. Such a situation is illustrated by the instances labelled in Figure 3.4.

Since the interest levels within the relevant community are heightened, being driven by the event(s), there is an opportunity for targeted interventions to make an impact within such a spiking phase. We shall introduce a model that addresses both the dynamic nature of the message-passing and the essentially static structure of the underlying 'who listens to whom' network across the relevant community.

Social influence was shown to play a crucial role in the propagation of information on Facebook [10]: 'Those who are exposed [to friends information] are significantly more likely to spread information and do so sooner than those who are not exposed.' Further empirical work appeared in Lin et al. (2013) [97] on major events during the 2012 US presidential election. The study found that human behaviour changes during a media activity, when information consumption is characterized by the availability of dual-screening technology (television and hand-held device) and real-time interaction. The authors proposed the term media event-driven behavioural change for this general effect, and showed that, for the data they collected, differences in behaviour were driven by the increasing attention given to a small cohort of elite users. We shall also focus on shared-attention, event-based setting, and the leadership role of some central users, but we shall also consider network behaviour.

The heightened full attention span setting proposed here is important as it implies that messages sent will be received more or less immediately. This is a problem with many dynamic (evolving network) analyses, since in general circumstances we may know when a tweet, SMS, message, or post is sent yet be unsure as to when (or even if) it was actually received. So not only is the current situation potentially actionable in real time, but it is also close to the best possible case of examining an active and interested community.

To be specific, let us consider Twitter activity, while noting that the same principles will apply to other time-dependent digital messaging systems. For the relevant set of n users, let A denote a corresponding binary $n \times n$ adjacency matrix, where $A_{ij} = 1$ if user i is known to receive and take notice of messages from user j. In this section, A may not be symmetric. Loosely, this might mean that i is known to be a Twitter follower of j, although in practice we have in mind the use of more concrete evidence that i cares about the tweets of j; for example via evidence from a history of retweets or mentions. For simplicity, we shall take a standard unit of time (one minute in the tests below) during which a user is assumed to send out at most one message.

We let $\mathbf{w}^{[k]} \in \{0, 1\}^n$ denote an indicator vector for the send activity at time interval $k = 1, 2, \ldots$, so that the ith component $\mathbf{w}_i^{[k]} = 1$ if user i tweeted in time interval k and is zero otherwise. Then set

$$\mathbf{q}^{[k]} = A\mathbf{w}^{[k]}$$

where $\mathbf{q}^{[k]} \in \mathbb{N}^n$ is such that the ith component counts how many messages were received by user i in this time interval.

Now we assume that the probabilistic behaviour of each individual (to either send a message or not) within the $(k + 1)$th time interval is conditional on the messages that he/she received within the kth time step:

$$P(\mathbf{w}_i^{[k+1]} = 1|\mathbf{w}^{[k]}, X) \propto \beta_i + \alpha \mathbf{q}_i.$$

Here, β_i is a basal rate that could be observed as individual i's spontaneous tweeting during eventless periods, or any other time. The second term on the right-hand side quantifies our assumption that, within the full attention span phase, activity is driven by a desire to join in with the current conversation and engage in topical 'banter'.

Without loss of generality we shall take the constant of proportionality to be unity: later we will discuss measurement and thus the scaling of $\beta = (\beta_1, \beta_2, \ldots, \beta_n)^T$), while α is still to be determined.

Rewriting the model in vector form and taking expected values we obtain the iteration:

$$\langle \mathbf{w}^{[k+1]} | \mathbf{w}^{[k]}, X \rangle = \beta + \alpha A \langle \mathbf{w}^{[k]} | X \rangle.$$

This type of iteration is familiar in many modelling and computation scenarios, notably in numerical analysis, and it is readily shown that as k increases, $\langle \mathbf{w}^{[k+1]} \rangle$ generically lines up along a preferred direction provided that the spectral radius of A is below $1/\alpha$. In this case, considering the limit $k \to \infty$, the resulting steady state value for $\langle \mathbf{w}^{[k]} \rangle$, which we denote by $\mathbf{w}^*$, satisfies

$$(I - \alpha A)\mathbf{w}^* = \beta.$$

To compute $\mathbf{w}^*$ requires the solution of a linear system involving the underlying network adjacency matrix, A. Because A is typically very sparse, this computation is feasible. We may regard $\mathbf{w}^*$ as a network centrality measure [108]. Indeed it is obviously closely related to the widely-used Katz centrality [86] introduced in Chapter 2 and especially Section 2.1.

We wish to use the expected values in $\mathbf{w}^*$ to predict a ranking of the users in order of their actual activity within some future event driven avalanche.

In any particular application we may proceed as follows. We define a 'business as usual' (BAU) period where the users operate at their basal rate, and a 'spike' period, where network activity has been dramatically increased by an external event. The basal tweet rate β_i for user i is taken to be their total number of BAU tweets. Since we are only concerned with the relative ranking induced by $\mathbf{w}^*$, there is no requirement to normalize this quantity. We also build the matrix A from BAU data, setting $A_{ij} = 1$ if i received at least one relevant tweet from j in this period, and zero otherwise.

We wish to judge the predictive power of $\mathbf{w}^*$. We shall do so by predicting the key users with $\mathbf{w}^*$ and recording their total Twitter activity during the spike period. In doing so we must address the matter of choosing α.

Our first example, taken from Higham et al. (2014) [77], uses data collected on 9 May 2013 surrounding the appointment of David Moyes as the new manager of Manchester United Football Club, following the retirement of Sir Alex Ferguson. This consisted of 298,335 time-stamped directed message-passing events involving 148,918 distinct Twitter accounts. The upper plot in Figure 3.5 shows the volume of tweets each minute. The largest spike in volume, at 486 minutes, corresponds to the official announcement of Moyes appointment. (The next largest peak, appearing earlier, corresponds to Everton

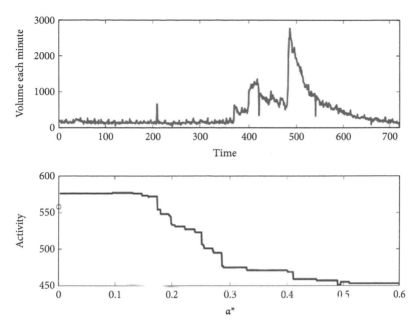

Fig 3.5 Upper: volume of tweets each minute for a conversation around Manchester United FC, 9 May 2013. Lower: activity of the predicted top one hundred tweeters during the spike phase, as a function of the response parameter, α^* [77].

FC announcing Moyes departure.) For the purposes of our test, we regard zero to 300 minutes as forming the BAU period where users operate at their basal rate. We define the spike period as lasting from the peak time of 486 minutes to the time of 541 minutes at which the activity level has decayed by a factor of four. The lower picture in Figure 3.5 shows the total spike period activity of the top one hundred ranked users, as a function of the response parameter, α. For compatibility with other tests, we present the results in terms of $\alpha^* = \alpha/\rho(A)$, where $\rho(A)$ denotes the spectral radius of A, so that $\alpha^* = 1$ becomes a natural (and practical) upper limit (where **w** becomes unbounded). The activity level for $\alpha^* = 0$ is marked with a circle. In this example, as soon as α^* increases beyond machine precision, the prediction improves (presumably by sorting out some dead heats in the BAU ranking). We also see that there is a broad range of lowish α^* values for which this improved prediction is obtained. There, **w*** does a good job in identifying the top tweeters *a priori*, based on only BAU data, held in (β, A).

Here is a second example from Higham et al. (2014) [77]. Figure 3.6 depicts results for a dual-screening conversation around an episode of the Channel

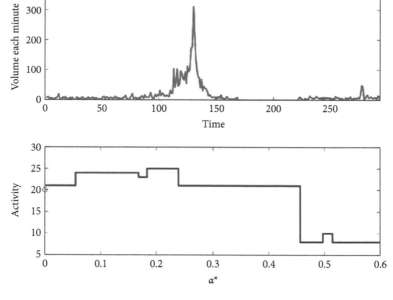

Fig 3.6 Upper: volume of tweets each minute for a conversation around a TV programme. Lower: activity of predicted top ten tweeters during the spike phase, as a function of the response parameter, α^* [77].

Four UK TV programme *Utopia*, involving 4,154 Twitter users. The spike at time 130 minutes corresponds to a particularly dramatic scene. We defined the spike to finish at 145 minutes, and took the BAU period to last from time zero to 120 minutes.

On the basis of these tests and others in Higham et al. (2014) [77], we may conclude that there is value in information about the underlying BAU static network structure, in predicting the top tweeters. In each case, incorporating a smallish value of the response parameter, α^*, of say 0.25, leads to an improvement.

3.6 Bootstrap Resampling and Sampling Errors

The central idea of bootstrap resampling is very straightforward [38], and it should be widely used. For data that arrives in real time (with velocity) and that may have avalanches, such as Twitter data, data from financial markets, or cascades of event-driven communications, some care must be taken to do the resampling (as explained below) as the data arrives, since we may need to take a live decision. So here, we illustrate a method of doing so.

Suppose that we have a set of data $Z = \{z_i | i = 1, \ldots, n\}$ and we calculate some statistic, θ, from Z. Then we shall write $\theta(Z)$. What is the sampling error on $\theta(Z)$? How far is it away from the result we would obtain if we had sampled the entire population of zs? Let us assume that in fact the elements of Z were sampled independently from some distribution $P(z|X)$. We do not know P. If we did, then we could make another data set, called Z', of the same size as the original by sampling independently from P. Then we could calculate $\theta(Z')$. By doing this repeatedly we could observe a distribution of values for θ. Then we could estimate a range of θ values that we could observe, for example some low and high percentiles (say 2.5% and 97.5%) to get a 95% confidence range. The beauty of this is that it works even when θ is very complicated and nonlinear.

In bootstrapping we do exactly this, except that since we do not know P we approximate it with a distribution of point masses over the discrete values held in the original set Z. Thus elements of each Z' are drawn independently from the n elements of Z. The reader is referred to a most readable account of this topic given in Efron and Tibshirani (1993) [38] by its discoverer.

When n is large, so large that Z cannot be held in available computer memory, we shall wind through Z reading in an observation at a time and we calculate some statistic θ by employing a collection of running sums or counters, denoted by C. This is usually the case with disaggregated shopping basket data from bricks-and-mortar or online retailers, for example, where we may be running over millions of shopping baskets, each containing a large number of items, and we keep some running-matrix-sums of instances or amounts of pairwise cross-purchasing, for many product items (called stock-keeping units, SKUs). Thus for each basket z_i we update all of our running sums and counts in C, and then move on to the next. Then finally we calculate θ from some possibly nonlinear operations over the collection C. This is also the case with Twitter data where we are taking unions of tweets (as they arrive) to form time-sliced adjacency matrices (as in a previous section).

In such a case, we have an outer loop over the n objects in Z. Now we wish to calculate an ensemble of values of θ for resampled data sets, Z', via other collections C'. Let us place an inner loop over K-ensemble of such collections, $C_j, j = 1, \ldots, K$.

Each time we read in a new observation z_i, we count it exactly once in C_1, the true collection of running sums and counts corresponding to Z, and a number of times $m_{i,j}$ in each of the ensemble of resampled collections C_j, for $j = 2, \ldots, K$. Thus z_i may not count at all towards some of the C_js (when $m_{i,j} = 0$), but it may contribute a number of times to other C_js. All that remains is to generate $m_{i,j}$. For each i and for each j it is to be drawn independently from a known distribution defined over the non-negative integers. This accounts for the number of times that each observations is resampled within each bootstrapped calculation of size n. So we have the probability that an observation is drawn randomly from a set n objects exactly $m \geq 0$ times out of n attempts:

$$P(m|n, X) = \frac{n!}{m!(n-m)!}(1/n)^m (1 - 1/n)^{n-m}.$$

In the limit for large n we have

$$P(m|\infty, X) = \frac{1}{m!} \lim_{n \to \infty} \frac{n!}{(n-m)!}(1/n)^m (1 - 1/n)^{n-m}.$$

Now

$$\lim_{n \to \infty} \frac{n!}{(n-m)!}(1/n)^m = \frac{n(n-1)\ldots(n-m+1)}{n^m} = 1.$$

Also

$$\lim_{n \to \infty} (1 - 1/n)^{n-m} = \lim_{n \to \infty} (1 - 1/n)^n \lim_{n \to \infty} (1 - 1/n)^{-k},$$

and the second limit tends to 1. The first limit is e^{-1} (see Exercise 3.6). Thus we have a Poisson distribution:

$$P(m|\infty, X) = \frac{e^{-1}}{m!}.$$

So as we read in each new observation, we may generate the resampling rates $m_{i,j}$ on the fly, from this Poisson approximation. Hence we can create a K-ensemble of collections of running counts and sums, C_j, in parallel, as required. Consequently bootstrapping is very much in play when n is very large. Of course the number of observations resampled from the original set Z and contributing to each C_j will contain approximately n observations. So care must be taken to count them if this will affect the consequent value of θ_j for that member of the ensemble. Take $K = 10^3$, sit back, and let the computer take the strain.

Bootstrapping the sampling errors of measures calculated from a captured data set (that is supposed to represent the entire distribution of possible behaviour) is very important when we present results to non-mathematicians.

Obviously when n is very large then even for nonlinear measures, θ, the sampling errors will hopefully be small. But care must be taken, because you may get down to some very small numbers of positive observations to be used to calculate θ very quickly.

The bootstrap works the data harder so as to produce a distribution of θ values, and thus estimate a sampling error on the original estimate. Conceptually this is what we have always done for simple statistics, such as moments. If we want to estimate a certain moment for a population from a sample, then we calculate a higher moment from that sample and use this to estimate the sampling error for the lower one.

Exercise 3.6

Show that $\lim_{n\to\infty}(1 - 1/n)^n = e^{-1}$.

Exercise 3.7

(a) What is the probability that an observation is included more than once in a resampled data set?

(b) What is the probability that the resampled data is exactly the same as the original sample?

When you look at shopping baskets and you ask about possible cross-purchasing of two products that each have an independent, basket penetration of 10^{-3}, then under the null hypothesis that they are purchased independently, we would expect to see them together in only one basket in a million. But with 10^7 customers with loyalty cards (like a very large supermarket chain), a whole year's shopping corresponds to say 10^9 baskets. So how many more than 1,000 baskets containing both products would be needed to make this cross-purchasing significant?

As an example application of bootstrapping, let us apply it to such *basket metrics*, a topic discussed in Solutions to exercise section. Suppose that we observe a very large number, n, of shopping baskets (or if we have loyalty cards then we may aggregate shopping by customer-fortnights, defining these as generalized *baskets*) and that we wish to determine the probability that buyers of one product also buy another in the same basket. Suppose that we restrict ourselves to K distinct products. Here, 'product' may mean an individual item (SKU) such as a single 300ml can of Diet Coke; or it may mean a specified group of products. Retailers such as hypermarkets and supermarkets may have many hundreds of subcategories, or be able to group products by brand (from a category hierarchy table).

As we read in each basket, we shall keep a running count of the number of baskets containing each product (and another of the total value of each basket containing each product, so we can say that product Y sells into one per cent of baskets but those account for twenty per cent of the retailer's total revenue, that is, it tends to appear in the biggest baskets). Let m_k denote the running count of the baskets, say $j = 1, \ldots, i$ that contain each product, $k = 1, \ldots, K$, starting at zero before any basket is observed. Let ι_{k_1, k_2} contain the similar running count of the number of the baskets $j = 1, \ldots, i$ containing both product k_1 and product k_2. Note the matrix of values, ι, is symmetric. Then once we read in the $i + 1$th basket we will observe a total of $q_{i+1} \in [1, K]$ distinct products all bought together within it, so we may update the corresponding q_i terms inside the m_ks, and update $q_{i+1}(q_{i+1} - 1)/2$ terms inside the upper triangle of the ι-matrix. At any point we can make the ι-matrix symmetric. Finally, when we have wound through all n baskets, we have two useful ways to normalize the ι-matrix. First, we can add one to the ι_{k_1, k_2} term and then divide this by $(m_{k_1} m_{k_2} + 2)$. This yields a similarity matrix, called the *cross-purchasing matrix*, that can be used to cluster products (see Chapter 1). If we add one to the ι_{k_1, k_2} term and then divide by $m_{k_1} + 2$, we obtain an estimate for the conditional probability:

$$P(\text{product } k_2 | \text{ product } k_1, X) = \frac{\iota_{k_1, k_2} + 1}{m_{k_1} + 2},$$

from Laplace's law of succession.

Of course, if these two products are in fact bought independently, then we should expect that this last is very close to an estimate for the basket penetration for product k_2:

$$P(\text{product } k_2 | \text{ product } k_1, \text{ products independent}, X) = \frac{m_{k_2} + 1}{n + 2}.$$

So we need to estimate a sampling errors on the quantities:

$$q_k = \frac{m_k}{n + 2} \text{ and } \kappa_{k_1, k_2} = \frac{\iota_{k_1, k_2} + 1}{m_{k_1} + 2}.$$

For the basket penetrations, q_k, the sampling error could be estimated directly, since the sample variance for the estimate q_k is

$$\frac{q_k(1 - q_k)}{n + 3},$$

see the Appendix. Yet the bootstrapping method will be more robust, especially if and when we get down to some relatively small numbers, and it can used to estimate ranges for each κ_{k_1,k_2} due to possible sampling errors. We may simply calculate multiple versions of the ι-matrix and the m_ks (within an inner loop) as we wind through the baskets (in an outer loop). The bootstrapping methodology also allows us the flexibility to estimate a sampling error range for anything else we care to calculate, from a variety of running counts or sums over the baskets, conditional on their contents.

3.7 Structure in Time Series

The search for structure within data, especially dynamical data, is a problem in many areas of applied mathematics, signals processing, and machine learning. These certainly qualify as 'big data' fields, for example analysis of climate data, industrial plant monitoring, or financial market data that are all in the form of multiple time series; or face recognition problems in the form of multiple image libraries (consisting of large matrices). However, they are of only peripheral interest in this book, as they are not immediately useful within customer-facing industries or behavioural analytics. In this chapter we have been mostly concerned with social networks, where the large number, n, is the number of the individual components (the vertices in networks). Yet within the engineering, image processing, or scientific fields, the large number is often the number of time points or the size of individual event data. Though both 'big', they are so in different dimensions, and the data are of a different shape. We know a little bit about a huge number of interacting components or we know a huge amount about a relatively small number of components.

When the data consist of a large time series (or multiple time series) there are some straightforward embedding and reconstruction questions to be answered. Assuming that the observed data is exhibited as a consequence of the long-term behaviour of an attractor for an underlying dynamical system, (a) can we reconstruct or estimate some properties of the attractor? (b) can we spot when some abnormal event is taking place? (c) can we use this knowledge to make some inferences? All of this needs to be done under the presence of noise and/or missing data.

Embedding theorems and methods are extremely useful in characterizing dynamics through an attempted reconstruction of the attractor (and its dynamics) within an estimated embedding space [18], usually this is called *state space embedding* or *singular spectrum analysis* [2]. The idea is to reconstruct an approximation to the underlying attractor by estimating a suitable embedding space to contain it, via an approximation to a lag correlation matrix. This requires us to make some assumptions about the nature of the noise present.

For a simple example of embedding, we start with a high resolution sequence $x_i = w_i + 1.5 \cos(i/\pi) + 2 \sin i$ $(i = 1, \ldots, 2000)$, where the w_i are randomly and independent drawn from a uniform distribution over $[-1,1]$. Then we partition the series so as to form a sequence of moving windows, in a trial dimension, or 'window length', in this case equal to 20:

$$W = (w_{ij}) \quad w_{ij} = x_{i+j-1}, i = 1, 1981, j = 1, \ldots, 20.$$

We wish to find projection of these window vectors (the rows of W) that contain as much as possible of the behaviour: that is, we wish to find orthogonal vectors $\mathbf{v}_k$ $(k = 1, 2, \ldots, 20)$ that successively maximize $\|W\mathbf{v}\|^2 = \mathbf{v}^T W^T W \mathbf{v}$. But the matrix $M = W^T W$ is real and self-adjoint, and this last is solved via the ordered eigenvalues and vectors of M, for preference using the singular value decomposition of W (as described in Chapter 1). In Figure 3.7 we show the spectrum of M and the resulting three-dimensional projection of W onto $W\mathbf{v}_j$. for $j = 1, 2, 3$. Clearly we recover a noisy winding map on the torus, corresponding to two non-commensurate oscillatory components within the signal. In Figure 3.8 we depict an example similar to that given originally in Broomhead and King (1986) [18] where a noisy (thirty per cent noise) chaotic signal, given by a solution the Lorenz system, shows that we may reconstruct a chaotic attractor even in the presence of random noise.

Given our interests in the spectral properties of similarity and adjacency matrices (in Chapter 1), and consequently in the singular value decomposition, SVD, its generalization to higher dimensional tensors is likely to be extremely useful as computational efficiency allow us to become more ambitious. This is a matter of current research and tool boxes are becoming available [88]. SVD representations of higher dimensional tensors are not unique though, so care needs to be taken. It is possible that these methods will become useful within topological data analysis (TDA) [19], but again any likely applications are extremely data hungry, and so this is most useful in fields of scientific interest,

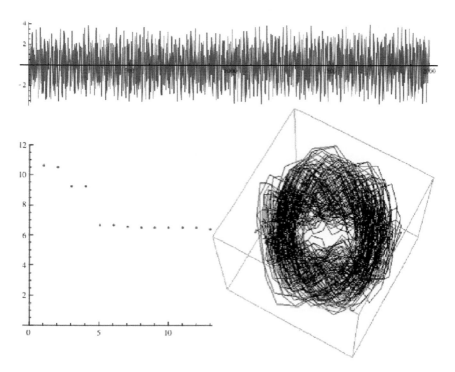

Fig 3.7 Example: two-cycle noisy time series, the log-eigenvalues of M, and the 3D embedding.

where large amounts are known about individual observations, rather than in social or commercial fields, where small amounts are known about a large number of observations.

3.8 Bayes Factors

In the last part of this chapter, we shall discuss briefly the problem of searching for drivers or indicators that might predict behavioural differences between individuals, before we have observed their actual behavioural data. Once we have classified a test population according to some detailed aspect of their behaviour, for example influencers versus non-influencers, we may wish to know if we could have inferred this earlier (*a priori*) from knowledge of other attributes (that may be causal, or else where both designations may be dependent on the same hidden factor). Suppose we have two sets containing different types of observations, generated under two alternative hypotheses, and that for each

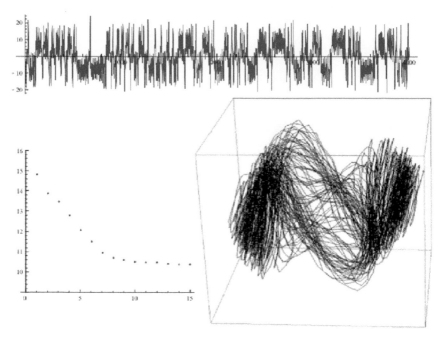

Fig 3.8 Example: solution of the Lorenz system with random noise added, the log-eigenvalues of M, and the 3D embedding.

observation we measure a large number of variables or attributes. Then we wish to know which of these attributes is the most useful in determining the set to which any new observation belongs. This is called supervised discrimination. Typically we will have a calibration set for which we know the type to which each observation belongs, and we wish to use these to find out how we can discriminate the type of any future observation based solely on the basis of a few of the attributes. A further discussion of classifiers for supervised discrimination is given in Chapter 5, and especially in Section 5.8.

If the observations are samples, then often we will have relatively few of these (100s in clinical tests or interviews), though we may have a large number of possible attributes with which to discriminate. If the attribute is age, we might split it into age bands. If the attributes are genomic, such as SNPs (typically trinomial variables: XY, XX, YY), then we may have thousands or hundreds of thousands available. Which of these attributes is a good marker for the type?

If the observations are shopper surveys then how can we know which households are likely to be big purchasers of certain cleaning products? Can we target households using knowledge of just a few well-chosen attributes and steal their purchasing from our competitors? This is called the 'golden question' problem: if I can only ask a householder just one or two questions from a large set of indirect test questions, then can I classify their level of consumption as either *small* or *large*?

Here we shall introduce a method that allows us to contrast different types of variable: binary flags, categorical variables, or real numbers, by taking an approach based on a Bayesian measure of the *evidence* that any attribute provides for distinguishing one set of classifications versus another. To begin with, we label one set of observations as 'cases', and the other as; 'controls', a nomenclature common in many scenarios. The 'evidence' in question is a data average of log likelihood ratios evaluated by Bayesian methods, each known as a log Bayes factor or 'LBF'.

Bayes factors arise when, given some new observation, E, we try to compare the probabilities of two mutually exclusive hypotheses, H_1 and H_2, that 'it is a case' and that 'it is a control', respectively.

As in the Appendix, it follows from Bayes that we have

$$\frac{P(H_1|E,X)}{P(H_2|E,X)} = \frac{P(E|H_1,X)}{P(E|H_2,X)} \cdot \frac{P(H_1|X)}{P(H_2|X)}. \tag{3.8}$$

As usual, we must also take account of any prior modelling information we have, X.

The first of the two right-hand terms is a ratio of the observation likelihood given each hypothesis, and is known as a Bayes factor (BF). The second ratio is the ratio of the priors of each hypothesis given what we 'know' (or can assert) already, called the prior odds. The value of Equation (3.8) is that it provides us with a route to quantitative calculation.

Calculating a Bayes factor is straightforward for us now: we have dealt with different types of models in the Appendix. If we consider an attribute that is real, z say, then we will select a probability distribution function to model $P(z|H_i,X)$. Given a calibration data set under each hypothesis, we will need to use these to tie down any parameters that are, 'hidden' in this notation, but that are involved in specifying this model. Suppose the model depends on some parameters, θ say, that we have selected our model for z (using our X!). For example θ might be the mean and variance of a normal distribution, or of log normal (if z must be positive); or an intensity for a Poisson process; and so on.

Let us write $P(z|H_i,X) = P(z|H_i,X)[\theta]$ to make the θ dependence explicit. Then we can use Bayes to calibrate the model. Let $f_0(\theta|X)$ be a prior probability density function for θ. Let D_i be a set of observations for z under H_i. Then we have the posterior for θ

$$f(\theta|D_i, X) = \prod_{z \in D_i} P(z|H_i, X)[\theta].f_0(\theta|X).$$

Now we must summarize this posterior, and select a good estimate for θ to use going forwards. Let us choose the value at the mode for simplicity.

Note that if f_0 is equal to one everywhere (and is improper), then selecting the mode is equivalent to 'maximum likelihood' estimation. We simply find the mode, the single value for θ that makes the observations in D_i most likely to have actually occurred. Hence we may estimate suitable θ-values to calibrate our models, using the case and control calibration data sets separately.

Alternatively, we might consider an attribute that is categorical , z say, with m mutually exclusive alternatives, then we will use a multinomial model and employ Laplace's law of succession so as to estimate the multinomial probabilities. Again we simply do this for each hypothesis, using case and control calibration data sets, respectively.

Of course we may also make hybrid models which are a combination of more than one measured attribute: for example, two trinomials together become a ninenomial; or a categorical and a real become a location model.

In a situation with many observations, if they are assumed independent, then an overall Bayes factor may be formed by multiplying all the individual ones together. This is equivalent to successive applications of (3.8). So let us assume independence.

Taking the logarithm of (3.8) is highly convenient at this point since the log of the product of separate BFs is simply the sum of the separate logarithms of each Bayes factor, called Log Bayes Factors or LBFs. The sum of the LBFs provides a measure of the total evidence that the observations offer in distinguishing one model from another.

In practice, each observation of a case or control may consist of one or more individual measured attribute $(Y_A, Y_B, \ldots)$. To manage this we consider each attribute in isolation from the others and evaluate an LBF for each single attribute separately. For an attribute Y, let y be a measured value observed from either a case or a control.

Then we define

$$LBF(y) = \log\left(\frac{P(y|H_1, X)}{P(y|H_2, X)}\right).$$

LBF(y) tells us the change in the log odds that *a single observation is a case*, as a result of knowing the attribute *Y* has value *y*. On the other hand −*LBF(y)* tells us the change in the log odds that the observation is a control, as a result of knowing the attribute *Y* has value *y*. The change in sign for the LBF is a consequence of the BF for a control hypothesis being the inverse of the BF for a case hypothesis (swap H_1 with H_2 in the BF ratio).

Let us consider a calibration data set, *E*, containing a total of *N* cases and controls, each with their corresponding measured *y*-values, say y_i for $i = 1, \ldots, N$. We will have a known indicator variable, with $r_i = 0$ for the cases and $r_i = 1$ for the controls. Suppose there are N_{ca} cases and N_{co} controls.

We will consider the odds on a hybrid hypothesis corresponding to *E*:

$$H = \cap_{i=1}^{N} h_i,$$

where each individual hypothesis refers to an individual observation:

$$h_i = \{ 'y_i \text{ is a case'} \text{ if } r_i = 0 \}, h_i = \{ 'y_i \text{ is a control'} \text{ if } r_i = 1 \}.$$

In fact we know that *H* is true for *E*. Let us see how knowledge of the *Y*-attribute, all of the *y*-values, might improve the odds.

There must of course be some prior odds for *H*. This uses no knowledges of the attribute *Y*, and it plays no role in the following.

Then we can calculate the posterior odds for *H* (assuming independence). We simply sum the LBFs (under each separate hypothesis for each corresponding observation). Summing the LBFs is equivalent to the product of the corresponding BFs. We have

$$\log O(H|E, X) = \sum_{i, \text{ Cases}} LBF(y_i) - \sum_{i, \text{ Controls}} LBF(y_i) + \log O(H|X).$$

Then consider the sum of 'signed LBFs' in this last formula:

$$\sum_{i=1}^{N} (-1)^{r_i} LBF(y_i) = \sum_{i, \text{ Cases}} LBF(y_i) - \sum_{i, \text{ Controls}} LBF(y_i).$$

This is the change in the log odds for the whole set of the correct hypotheses, *H*, as a result of observing the set of y_is.

Since for *E* we actually know that *H* is true, we wish to find an attribute for which this sum is as large as possible.

Finally, since we wish to compare this term across different attributes (comparing apples with oranges, reals with categoricals, etc), we must take care to make sure the evidence is fair to all. Often for practical reasons some measurements are not available in all observations. To allow fair comparison between attributes we average each 'signed LBF' contribution over its available number of observations.

This gives us a data normalized value as a final evidential value, *ev*:

$$ev(Y) = \frac{1}{N} \sum_{i=1}^{N} (-1)^{r_i} LBF(y_i). \tag{3.9}$$

Remark: the observed arithmetic mean of the signed LBFs is equivalent to the log of the geometric mean of the corresponding BFs. It is thus the increase in the log odds, $O(H|X)$ expected from a single measurement of the chosen attribute.

The normalized evidence goes a fair way to compare the discriminative power of different attributes, but it does not indicate the significance of each on its own. To assess this, we may evaluate the evidence from the data after 'maliciously' shuffling the measured values of an attribute amongst its available case plus control data. After each shuffle we get evidence that is only coincidental. We can then assess the significance of the original evidence by how it compares with the highest coincidental evidence from a large number of shuffles. For example, if the genuine evidence is 2.5 and the coincidental evidence from 200 shuffles is less than 2.5 on more than 190 occasions, then it is clear that the genuine evidence is significant with approximately 95% confidence.

See the remarks 'On resampling and shuffling' at the end of this chapter.

Exercise 3.8

We have one hundred cases—individuals who are each amongst the top influencers within some topic-based social network (H_1), and eighty controls (H_2) who are not amongst the top influencers. We also have an attribute called 'gender' with values 0 and 1. Suppose that sixty of the cases are 0s and forty are 1s, while thirty of the controls are 0s and fifty are 1s. Show that

$$ev(\text{Gender}) = \frac{60\log(8/5) + 40\log(16/25) - 30\log(8/5) - 50\log(16/25)}{180}$$

$$= \frac{30\log(8/5) - 10\log(16/25)}{180} = 0.103.$$

Now suppose we also have another attribute called 'County' with three values: Oxon, Berks, and Bucks (Oxon, Berks, and Bucks are three nearby English counties where these individual cases and controls all reside). Suppose fifty of the cases are Oxon, thirty are Berks, and twenty are Bucks; while thirty of the controls are Oxon, thirty are Berks, and twenty are Bucks.

Is the 'county' attribute more or less useful than the 'gender' attribute in inferring any individual is a case?

A Personal View

On the Usefulness of Examples

I was tutored as an undergraduate in mathematics at Bristol by John P. Cleave, a logician. A few years later I met him again when I was an external industrial advisor to the engineering mathematics activities at Bristol, and a number of academics from different schools in science and engineering had got together to discuss risk, uncertainty, and logic, and a possible multidisciplinary activity. There was a good intellectual kick-around until people started to get specific about a planned research proposal. Above the argument John cried 'Stop! Please don't confuse me with an example!'

This needs to be remembered. Examples are supposed to shed some light into dark corners and to illustrate, or to be used as benchmarks for practice. They do not add weight. Our primary aim is first to capture and master the concept at hand then, at some point when we have it, we are ready to abstract and shift sideways. More examples are not necessarily better. Even an encyclopaedia full of well-worked examples is not as interesting or exciting as the next radical leap forward.

On Social Media

We are all influenced in ways that we are not consciously aware of. The power of advertising to seduce us into making choices or adopting certain positions, and the success of grooming in capturing recruits to radical causes, or for purposes of exploitation, are magnified by social media and digital platforms. Why should this be so?

As mathematicians we are challenged to replicate these effects, and thus to forecast them. Humans do not make decisions rationally, by seeking out and fully

continued

On Social Media *Continued*

considering all necessary attributes of the problem, and thus maximizing some utility or benefit, but by using *heuristics*. These are simplified thought processes, using a limited amount, or incomplete set, of data, and they are at the root of behavioural economics and prospect theory. In his book *Thinking, Fast and Slow*, Daniel Kahneman [84] shared a lifetime of experience, and reading that book caused an epiphany for me. He set out how humans reason subconsciously so as to make decisions in real-time that turn out to be mostly good ones. But these heuristics may also produce cognitive illusions, and we are easily tricked (see also Ariely [6]). It is a useful working hypothesis that the impact of heuristics is especially important and amplified online, where the default persona adopted by most folk is far more open than within our real lives (when we are in rooms full of strangers, for example), and access to others' opinions and behaviours is made easy. In any analysis of online influencing (from, say, micro blogs and Twitter) we can examine sentiment, and we can amplify the sentiments expressed via broadcast centrality of the influencers within conversations. Thus we can see who is directionally influencing various opposing opinions or ideas. Good.

Yet there is today a lack of a mathematical framework for the usage and impacts of heuristics. How do people respond to influences and information? How do the fast-thinking subconscious processes control their subsequent behaviour? If such a science were available, we could better understand how terrorist organizations can successfully tap into these (intentionally or not). We might also better understand successful digital marketing. Marketeers already know that the medium is the message, not the apparent message content. That is why many highly successful campaigns do not contain any specifics. The whole medium communicates the feeling you are supposed to have. In Heath (2012) [72] the author explains this with many examples from UK television advertisements. We need an equivalent summary for online marketing experience. We urgently need both concepts and mathematical models. This field represents some real challenge with implications both for commercial marketing and for security and defence activities.

On Resampling and Shuffling

In certain applications, one constructs an evolving network in discrete time $\{A_k \,|\, k = 1, \ldots, K\}$ from the rush of emails, texts, tweets. Sometimes it is good practice to limit the network to a predefined or prequalified set of targets; sometimes it is not. In any case, suppose we extract the sequence and calculate Q as designed in Chapter 2. One might make a scatter plot of the broadcasts centralities (the row sums of Q) against the receive centralities (the column sums of Q). Immediately we see some anomalies, for example those individuals that have a high broadcast centralities and low receive centralities. We know that the asymmetries arise of size $O(\alpha^2)$ from the ordered products pairs of adjacency matrices (see Exercise 2.5 in Chapter 2). Suppose that we suspect that we have measured nothing special: if n is very large then surely we would expect to find some extreme individuals occurring at random.

A computationally simple way around this question is to assume a null hypothesis that there is no dynamic behaviour held within $\{A_k \,|k = 1, \ldots, K\}$, as measured by $\mathcal{Q}$, beyond that observable for a similar random sequence of such adjacency matrices.

To test this, suppose that we form a new evolving network by shuffling (randomly permuting) the original sequence $\{A_{q(k)}|k = 1, \ldots, K\}$ (here q permutes the index, mapping the set $\{1, 2, \ldots, K\}$ onto itself). Then all of the non-time dependent statistics of the sequence (oblivious to the true sequential ordering) remain the same. There are the same number of edges and each vertex has the same total degree. Then we may calculate the Katz centrality matrix, $\mathcal{Q}_q$, for this time-shuffled evolving network. Clearly, if we are claiming to have observed some significant behaviour in the real sequence, measured by a statistic $\theta(\mathcal{Q})$ say, then that value must be an extreme when compared to a distribution of such values, obtainable by $\theta(\mathcal{Q}_q)$, under the null hypothesis. It is easy to find this distribution by repeatedly permuting the sequence and recalculating $\mathcal{Q}_q$ and thus $\theta(\mathcal{Q}_q)$. We can express the result as a p-value, giving the probability of $\theta(\mathcal{Q}_q)$, under the null hypothesis, exceeding the true value $\theta(\mathcal{Q})$.

One of the first times I needed to do this was in some research on the analysis of human brains. We analysed tiny voxels (volumes) within a human brain and their pairwise interactions as if the whole was a complex social network. The observed behaviour was time dependent and provided by the blood oxygen level within each voxel (a surrogate for activity and energy used). The problem was raised by collaborators from psychology as to whether the timescales measured by fMRI scan, which are relatively long scales in neurological terms, were actually able to detect anything at all of interest to them. For θ we chose the mean absolute difference in broadcast and receive centralities ($\|\mathcal{Q}\mathbf{s} - \mathcal{Q}^T\mathbf{s}\|_1/n$), while n was many hundreds of thousands. The results were quite stark: the asymmetries observed within the true data were extremely significant and we could reject the null hypothesis with a very high degree of confidence, $1 - p$. Whatever we were measuring, it was certainly not present within the sequences generated under the null hypothesis.

Shuffling is drawing without repetition, whereas in bootstrap sampling we usually draw with repetitions (hence the Poisson approximation when the number of objects in the sample is large). Such methods are indispensable for two reasons. First, it is an honourable thing to let the computer do our work for us (we are in

continued

On Resampling and Shuffling *Continued*

analytics strategy after all, while the machine is in operations). Second, this type of demonstration is very easily understood by non-technical people. We have shown that if we randomize the sequence, we cannot observe the behaviour that we have identified. In many companies we rely on a bridge of trust between ourselves, as analysts, and those who must take action consequent to our insights. This type of method helps cement that trust.

CHAPTER 4

Clustering and unsupervised classification

In this chapter we consider one of the most basic questions for analytics, especially from within customer-facing industries: what kind of distinct groups of customers do we have? A business-driven approach might assert or guess the answer to this question, and get on with measuring customer *segments*, which is the usual analytics term for distinct customer groups defined by observable attributes. An analytics-driven approach should be to let the data speak for itself and not to assume that we can second guess the answer.

As customer data has both volume and velocity (the rate of new transactions), it is increasingly desirable to seek a set of segments that are behaviour-based and are thus dynamic; and are not defined in terms of static attributes such as location, socio-economic class, income, family size, total spend, and so on. These behavioural definitions may be applied to resegment the customer base every few weeks, based on each individual's most recent behavioural attributes, and then the outputs can reflect the pace of change of various customer-facing activities.

There are a number of mathematical ideas and methods to consider. Yet these apply much more widely to a general class of unsupervised discrimination and incomplete data problems.

Mathematical Underpinnings of Analytics. First Edition. Peter Grindrod.
© Peter Grindrod 2015. Published in 2015 by Oxford University Press.

4.1 Simple *K*-means Clustering

Here, we consider the problem of partitioning a set of n observations, as-sumed independent, into a discrete set of *clusters* (subsets of observations that are more similar than observations drown from different sets). In this well-known simple method, one assumes that each observation is just a vector of real numbers: $\mathbf{x}_i \in \mathbb{R}^m$ $i = 1, \ldots, n$.

We think of each observation as an independent random choice drawn from one of a number, K, of (hidden, initially unknown) classes, Y_k for $k = 1, \ldots, K$. In K-means these classes are defined as subsets of $\mathbf{x}$-space ($\mathbb{R}^m$) that form a proper partition. Yet the information about the classes is hidden from us. So the data is *incomplete*, and we must infer it along with the definitions of the classes themselves.

In the K-means method of clustering we shall further assume that each class, Y_k, is completely defined by a specific vector, called the 'centroid' and denoted by $\mathbf{x}^{[k]}$. Then $\mathbf{x}_i \in Y_{k^*}$ if, and only if, $k^* \in \{1, \ldots, K\}$ denotes the class that minimizes the distance to the corresponding centroid, $||\mathbf{x}_i - \mathbf{x}^{[k]}||$, in the usual Euclidean norm. We write

$$k^* = \arg\min_k ||\mathbf{x}_i - \mathbf{x}^{[k]}||.$$

Thus if the centroids are known, it is easy to attribute all of the observations.

Conversely if we have all of n present attributions, and a subset of the ob-servations are attributed to a particular class, Y_k say, then we could make an *a-posteriori* estimate for the corresponding centroid to be the observed mean of the present exemplary members:

$$\hat{\mathbf{x}}^{[k]} = \text{mean}\{\mathbf{x} \in Y_k\}.$$

In the K-means algorithm we simply iterate these two processes. We allocate the observations to their most likely classes, and then we readjust the centroids, defining each class, using the estimate from the means of the partitioned data. Then we repeat this process until the centroids remain constant at two succes-sive iterates, so that no more observations would ever be reclassified, and thus the classes would remain fixed if we continued.

To start we may generate random centroids (by choosing K of the observa-tions at random, for example).

4.2 Finite Mixture Models and the EM Algorithm

Assume that a single observation consists of a number of real and categorical variables, and that we have n observations in total. Each observation $\mathbf{x}_i$ for $i = 1, \ldots, n$ lives in a well-defined *feature space*, called $\mathbf{x}$-space.

We model each observation as a random choice drawn from one of a number, K, of (hidden, initially unknown) classes Y_k for $k = 1, \ldots, K$. For each class we have a probability distribution defined over $\mathbf{x}$-space, say $P(\mathbf{x}|\lambda_k, X) = F(\mathbf{x}, \lambda_k)$. Here λ_k must be in some subset Λ of $\mathbb{R}^m$, say, and is a vector of parameters that together completely specify the distribution via some given (assumed) function F. And X, as usual, represents 'everything else that we know' *a priori*.

For all possible values of λ fixed in Λ we must have the probability $F(\mathbf{x}, \lambda) \geq 0$ and it must sum to unity as $\mathbf{x}$ varies over $\mathbf{x}$-space. In this framework, the distributions for each class, k, are all of the same type: they merely differ through their particular defining parameter values, λ_k.

Note that typically some of the attributes observed within $\mathbf{x}$ will be categorical variables and that particular dimension might be parameterized as a multinomial model. Some attributes might be real numbers, and distributed by Gaussian, log-normal, Dirichelet or Poisson distributions, etc all suitable when parameterized. Some might be hybrid, or have multiple levels, with the distribution for an attribute at a lower level being conditional on a higher level categorical (called an 'indicator') variable; and so on. In any case, each class Y_k both defines and results in a distribution $F(\mathbf{x}, \lambda_k)$ over $\mathbf{x}$-space, completely parameterized by the vector $\lambda_k \in \Lambda$ in $\mathbb{R}^m$.

Let H_k denote the hypothesis that a random observation is drawn from class Y_k, for each $k = 1, \ldots, K$. A *finite mixture model* represents the whole set of N observations as a finite mixture of independent observations drawn from across the K classes. The model comprises the mixture probabilities (the prior probabilities that H_k is true for a random observation):

$$\pi_k = P(H_k|X) \quad k = 1, \ldots, K,$$

which sum to unity; and the class parameters

$$\lambda_k \quad k = 1, \ldots, K,$$

which define the conditional distributions via the functional form of F.

The full distribution for all observations over $\mathbf{x}$-space is given by the mixture sum

$$\sum_{k=1}^{K} \pi_k F(\mathbf{x}, \lambda_k).$$

It is from this that our observations are assumed to have been drawn.

In fact, each observation $\mathbf{x}_i$ is incomplete because there is a latent, or hidden, variable, say $z_i \in \{1, \ldots, K\}$, indicating the class k that actually generated the corresponding observation.

Our task is to estimate the π_ks and the λ_ks, given the set of actual observations, and in doing so, we will make some assertion about the likely value of the z_is.

The likelihood of the observed data, assumed independent, conditional on the observed and unobserved values, and on the parameters, is thus

$$\mathcal{L} = \prod_{i=1}^{n} \pi_{z_i} F(\mathbf{x}_i, \lambda_{z_i}). \tag{4.1}$$

One possible strategy is as follows. If we knew the value of the class parameters and the mixing probabilities, $\{\lambda_k, \pi_k\}$, then we might find the values of the latent variables, $\{z_i\}$, by maximizing the log-likelihood, somehow searching over all possible combinations. Conversely, if we knew the values of all of the latent variables, $\{z_i\}$, then we might find an estimate for the class parameters and the mixing probabilities, $\{\lambda_k, \pi_k\}$, fairly easily: typically by calibrating (by maximum likelihood estimation) each of the individual class distributions independently.

So we might iterate:

(a) initialize the class parameters and the mixing probabilities, $\{\lambda_k, \pi_k\}$, to some random values;

(b) compute the best value for the latent variables, $\{z_i\}$, given those parameter values and the mixing probabilities;

(c) employ the latent variables found in (b) to find better estimates for the class parameters and the mixing probabilities.

Repeat steps (b) and (c).

Step (b) is called the "expectation" step, while step (c) is called the maximization step. Together they form a type of Expectation-Maximisation algorithm, or EM for short.

This algorithm is commonly called *hard EM* since the discrete values for the latent variables are estimated in each step (b). The K-means algorithm in the last section is an example of such an algorithm.

However, in this section we shall be much more flexible. The problem with hard *EM* is that there may well be large parts of **x**-space where it is a rather close call as to which of several classes might generate such an observation (the $\pi_k F(\mathbf{x}|\lambda_k)$ being similar for a subset of the ks). So *quenching* to a single most likely class may well be grossly unfair on some of the other classes. Therefore instead of making such a hard choice for the latent variables (given the current parameter values), we may instead determine the probability of each possible value of $z_i \in \{1, \dots, K\}$ for each data observation, and then use those probabilities as weights in subsequently recalibrating the parameters for each class. In essence this replaces a single observation (with a hard latent variable z_k) with a set of observations having an identical **x** distributed according across the classes within the mixture. The resulting algorithm is sometimes called the *soft EM* algorithm, or more commonly the EM algorithm. Here we follow McLachlan and Peel (2000) [101].

Suppose first that we have a set of n observations, the ith one of which we presently believe to be have been independently drawn under hypothesis H_k, with probability $\tau_{ik} \in (0, 1)$. That is, τ_{ik} denotes our (soft) estimate for the probability that $z_i = k$: that observation i actually emanated from class k. Notice that $\sum_{k=1}^{K} \tau_{ik} = 1$, since we shall assume that it must have emanated from one or other of the K classes.

Then, assuming that all the soft latent variables, the τs, and the parameters, the λs and the πs, are known, then the likelihood of observing this data, $\{\mathbf{x}_i\}$, given previously by (4.1), becomes

$$\mathcal{L} \equiv P(\{\mathbf{x}_i\}|\{\lambda_k, \pi_k\}, \{\tau_{ik}\}, X) = \prod_{i=1}^{n} \prod_{k=1}^{K} (\pi_k F(\mathbf{x}_i, \lambda_k))^{\tau_{ik}}. \tag{4.2}$$

We have split the single ith observation in (4.1) into K observations each appropriately weighted by τ_{ik}. If ever the $\tau_{ik} = \delta_{k,z_i}$ for each k (and we know exactly which class generated each observation) then (4.2) becomes (4.1).

It is theoretically most convenient (and will avoid any small number issues in practice) if we take logs in (4.2) and consider the log-likelihood of observing that data:

$$\log \mathcal{L} \equiv \sum_{i=1}^{n} \sum_{k=1}^{K} \tau_{ik} \log (\pi_k F(\mathbf{x}_i, \lambda_k)) = \sum_{i=1}^{n} \sum_{k=1}^{K} \tau_{ik} \left(\log F(\mathbf{x}_i, \lambda_k) + \log \pi_k \right).$$

$$\tag{4.3}$$

Now assume the soft latent variables (probabilistic attributions of observations to the classes), $\{\tau_{ik}, i = 1, \ldots, n, k = 1, \ldots, K\}$, are all known but that the parameters $\{(\lambda_k, \pi_k), k = 1, \ldots, K\}$ are to be determined.

Then we must choose $\lambda_k \in \Lambda$ and the π_k so as to maximize $\log \mathcal{L}$ in (4.3).

Fixing k, for each λ_k this requires we maximize the simpler form

$$\sum_{i=1}^{n} \tau_{ik} \log F(\mathbf{x}_i, \lambda_k). \tag{4.4}$$

This will depend on the form of F, and is addressed analytically below in a few specific cases. In general this might be achieved numerically using a gradient search method or non-derivative method.

Similarly we may estimate $\{\pi_k | k = 1, \ldots, K\}$, so as to maximize (4.3) while constrained to sum to unity by

$$\pi_k = \frac{1}{n} \sum_{i=1}^{n} \tau_{ik}. \tag{4.5}$$

Exercise 4.1

Show that (4.5) holds at a maximum of $\log \mathcal{L}$ in (4.3) by using a Lagrange multiplier.

This, optimization process is called the 'M step', (M for maximization), where we locate and successively relocate the model-defining probabilities and parameters, $\{(\lambda_k, \pi_k), k = 1, \ldots, K\}$, so as to maximize $\log \mathcal{L}$, given our probabilistic attributions of observations to the classes, $\{\tau_{ik}, i = 1, \ldots, n, k = 1, \ldots, K\}$. Recall that $\tau_{ik} = P(\mathbf{x}_i \text{ generated under } H_k | X)$.

Now suppose the completely opposite situation occurs. Suppose that we have access to the values for all of the parameters and mixing probabilities, $\{(\lambda_k, \pi_k), k = 1, \ldots, K\}$, and that we need to estimate the probability that each observation has actually been generated by each one of the possible K classes: $\{\tau_{ik}, i = 1, \ldots, n, k = 1, \ldots, K\}$.

Given $\mathbf{x}_i$, say, we write the posterior

$$P(\mathbf{x}_i \text{ is generated under } H_k | \{(\lambda_k, \pi_k), k = 1, \ldots, K\}, X) = \frac{\pi_k F(\mathbf{x}_i, \lambda_k)}{\sum_{k'=1}^{K} \pi_{k'} F(\mathbf{x}_i, \lambda_{k'})}.$$

But this is precisely the form required to update our estimates for the τ_{ik}s.

This probabilistic attribution process is called the 'E step', (E for expectation).

So again, we have a two-step process. We may start from random seeds and iterate.

The devil here really is in the detail of the types of distribution that are to be used and their consequent parameterizations. It is very common to find EM codes that allow Gaussian or other well-used classes of distribution, since the maximum likelihood calibration can usually be reduced to algebra in such cases. We pause briefly to illustrate the use of Gaussian (normal) distributions (that of course produce quadratic forms to be optimized) and multinomial distributions, for real and categorical components of $\mathbf{x}$, respectively.

For example, suppose $\mathbf{x}$ contains a real valued component x, say, that is to be distributed by a Gaussian and is uncorrelated with any of the other components, then we have

$$F(\mathbf{x}, \lambda) - \frac{e^{-(x-\mu)^2/2\sigma^2}}{\sigma\sqrt{2\pi}} \tilde{F}(\tilde{\mathbf{x}}, \tilde{\lambda}),$$

where $\tilde{F}$ is some function of $\tilde{\mathbf{x}} = \mathbf{x}\backslash x$ ($\mathbf{x}$ sans x) and $\tilde{\lambda} = \lambda\backslash(\mu, \sigma)$.

Now consider (4.4), where we have fixed k and wish to find estimates for λ_k. We substitute for $F(\mathbf{x}_i, \lambda_k)$ and maximize with respect to μ_k and σ_k, the mean and standard deviation of the normal distribution for x corresponding to the kth class (independently of the other parameters held in $\tilde{\lambda}_k$).

We seek to maximize the following, with respect to (μ_k, σ_k):

$$\sum_{i=1}^{n} \tau_{ik} \left(-(x_i - \mu_k)^2/2\sigma_k^2 + \log \tilde{F}(\tilde{\mathbf{x}}_i, \tilde{\lambda}_k) \right) - n \log \sigma_k - (n/2) \log(2\pi).$$

This is the usual weighted quadratic optimization problem with solution

$$\mu_k = \sum_{i=1}^{n} \tau_{ik} x_i, \quad \sigma_k^2 = (1/n) \sum_{i=1}^{n} \tau_{ik}(x_i - \mu_k)^2.$$

So Gaussian distributions (and indeed multivariate Gaussians) are dealt with in the time-honoured manner. Note that if all of the components of $\mathbf{x}$ are real and independently distributed by Gaussians, then the EM method here results in a more sophisticated version of K-means, since the distributions instigates distinct data-driven weights (set by the variances) for each component (dimension) and for each class. If all observations are finally allocated to their most likely (modal) class, then the result is a partition which is in a much more general class than that achievable by centroid-defined clusters in any globally weighted Euclidean norm.

Often more useful are models that allow for **x** to have some categorical components, but this feature is commonly unavailable in standard packages. Again, one may calibrate the consequent multinomial distributions from weighted observations as follows.

Suppose **x** contains a categorical valued component, x say, with $R \geq 2$ categories that is independent from other components of **x** and to be distributed by a multinomial distribution with parameters $(q_1, q_2, \ldots, q_R) \in [0, 1]^r$ that sum to unity. These correspond to probabilities that an observation is in one or other of the R categories, say $\{C_1, C_2, \ldots, C_R\}$. For the kth class ($k = 1, \ldots, K$) these probabilities will be denoted by $(q_{k,1}, \ldots, q_{k,R})$, and these parameters are all contained within the vector λ_k. Then, proceeding just as before, from (4.4), for each $k = 1, \ldots, K$, we should choose the $(q_{k,1}, \ldots, q_{k,R})$ so as to maximize

$$\sum_{i=1}^{n} \tau_{ik} \log q_{k,r_i}$$

where the categorical component of the ith observation, x_i, is in C_{r_i}. So

$$q_{k,r} = \sum_{x_i \in C_r} \tau_{ki}/\eta \quad r = 1, 2, \ldots, R.$$

where η is the normalizing constant $\sum_{i=1}^{n} \tau_{ki}$.

Hence multinomial distributions are also dealt with in a time-honoured manner; though note the maximum likelihood estimate selects the mode, not via Laplace's law of succession.

Computationally, for both Gaussian and multinomial distributions, this process is highly efficient because every time we ever use this estimate, for all K classes and at any iteration, all that ever changes are the present (assumed) values for the τ_{ik} that act as weights. So a computer implication does the same dreary thing a huge number of times: a candidate for parallelization if ever we saw one.

Returning to the more general setting, in some given applications within industry one might have prior knowledge about the very specific nature of the distributions achieved by different classes. This is contained within $X=$'everything that we as modellers know *a priori*'. And indeed we may have some requirements or constraints around the numbers and types of classes:

there is no compulsion that all of the classes be recalibrations of a common model. So one may very well have to code-up a bespoke version of the EM algorithm for proprietary purposes. This should be straightforward now, given the examples exhibited previously.

We discuss some such commercial examples in Sections 4.3 to 4.6.

4.3 Practicalities

One should never blindly apply a technique such as EM, either off-the-shelf or by coding -up you own version (with your favourite distributions).

In the first place for bigdata, there is an issue as to what gets held in memory. If n is huge then the **x**s may not be held in memory. One could take a largish random sub-sample to establish a finite mixture model, or else take care that the implementation simply winds through the observations one at a time, calculating the running sums in the maximum likelihood parameters for all classes (so the outer 'loop' is around the observations which can be sequentially read in at every iteration). Note that the τs will need to be dealt with similarly (by being written out and read back in).

From a modelling perspective, it is important to eyeball the data before commencing with anything. We are modelling the distributions of each component of **x** so it makes sense to look at the observed distributions. If components are observed to be correlated (check on a sub-sample), our choices should reflect this. More importantly, just because we are presented with observable measurements as components we do not have to exploit them as they are. For example, if a component is positive and real, it may be best to take logs before distributing it normally (or equivalently invoke a log-normal distribution). Other types of distribution, such as exponential, fat-tailed (Dirichlet), and so on could be coded-up and made available.

A word about components of **x** that are integer counts is in order here. These are very common in applications within customer-facing businesses where we may count instances of a customer exhibiting different types of transaction or contact over some fixed time period. There are big differences between a count of zero (the action was never observed in the time period), a count of one (observed just one time), and a count of two. But how different is a count of thirty-one as opposed to thirty-two? Rather than crudely treat the integer count as a real or as a categorical, it is a good idea to map the counts onto categories that grow on a log scale. For example, if j is the count then this can be mapped to an integer ceiling, say $g(j) = [\log_2(1 + j)]^+$. So then $0 \to 0$, $1 \to 1$, $\{2, 3\} \to 2$, $\{4, 5, 6, 7\} \to 3$, and so on up to some maximum class indicator. Eyeballing such distributions both before and after transformation is absolutely essential.

4.4 Behavioural Segmentation of Energy Demand Profiles

Here we employ the EM framework to specify thirty micro behavioural states within the weekly patterns of domestic customer electricity usage, observed in thirty minute smart meter data profiles. (Typically these would be useful for operational purposes: targeting communications and monitoring trends and changes in behaviour.) If necessary these could be combined to form, say, seven to ten macro states that would useful for management reporting and strategic purposes.

The goal of the analytics is to employ smart meter data to discover a segmentation of customers based solely on their actual behaviour over a short time period, and not on their assets or their charging plans. For this project we choose the time period to be a week. This is the minimum time period over which credible usage patterns might be established, since many customers show different behaviours on specific days of the week (weekend versus weekday differences, for example).

The data consists of smart meter readings of energy usage over thirty minute periods for customers, running over a year or so. We considered 111,669 complete customer weeks (CWs), with each CW containing 336 consecutive readings of the energy used over each thirty minute period.

We first define a set of metrics representing attributes that capture behaviour of potential interest from the raw data. These will be used to represent each CW as a point in attribute space.

We shall employ the following eighteen features listed and defined in Table 4.1. They all require a full week of smart meter readings, but apply over different timescales as noted. Some of these attributes use the concept that usage at any time was habitual 'baseline' usage (representing everyday usage) and additional above baseline consumption (representing discretionary or occasional usage). Accordingly a set of forty-eight daily habitual, or 'baseline' values, are defined for each CW, one for each thirty minute period of a day.

Table 4.1 *Weekly demand attributes*

Name	Definition	Units
av-daily	Average of daily usage	KWh
var-daily	Variance of daily usage	KWh^2
av-baseline	Average of daily baseline	KWh
var-baseline	Variance of daily baseline	KWh^2
av-abovebase-weekday	Average of daily usage above baseline Mon–Fri	KWh
var-abovebase-weekday	Variance of daily usage above baseline Mon–Fri	KWh^2
av-abovebase-weekendday	Average of daily usage above baseline Sat–Sun	KWh
var-abovebase-weekendday	Variance of daily usage above baseline Sat–Sun	KWh^2
av-abovebase-daytime	Average of daily usage above baseline 07:00–15:30	KWh
var-abovebase-daytime	Variance of daily usage above baseline 07:00–15:30	KWh^2
av-abovebase-evening	Average of daily usage above baseline 16:00–23:30	KWh
var-abovebase-evening	Variance of daily usage above baseline 16:00–23:30	KWh^2
av-abovebase-offpeak	Average of daily usage above baseline 01:00–06:30	KWh
var-abovebase-offpeak	Variance of daily usage above baseline 01:00–06:30	KWh^2
av-baseline-daytime	Average of baseline usage 07:00–1530	KWh
av-baseline-evening	Average of baseline usage 16:00–23:30	KWh
av-baseline-offpeak	Average of baseline usage 01:00–06:30	KWh
av-dailysmoothness	Average of daily smoothness within a week	–

These values are derived by taking the smallest meter reading in each thirty minute period of the day throughout the seven days of the week. The concept is illustrated in Figure 4.1.

The feature av-baseline is the average of the daily baseline values. All 'above baseline' features are derived by subtracting the baseline values from actual use.

A measure of the smoothness within any day is the log-correlation between successive thirty min usage values throughout the day. The av-dailysmoothness is an average of those daily smoothness values over the seven days of the week.

The population of CWs in the eighteen-dimensional feature space will be modelled as a mixture of thirty states. This is quite a high resolution segmentation, but will allow tactical analysis and operational activity to take place on small two to five per cent segments. In practice, we might look at different

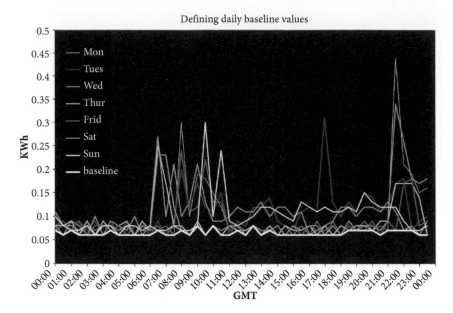

Fig 4.1 The daily baseline concept

segmentations with higher and lower numbers of classes and select a solution—such as the one presented here—for its relative uniformity of state sizes.

The distribution of each micro-state sub-population is modelled by the EM algorithm as a product of eighteen normal distributions, each having an unknown mean and standard deviation. In all this implies 1,109 unknown parameters (since mixing parameters, the πs, have twenty-nine degrees of freedom).

Visualization for presentation and discussion is obviously an issue, as there are many possible two-dimensional projections, but fortunately a great deal of insight may be gleaned from a few of them. In Figures 4.2 and 4.3 we use the projections of a 30,000 random sub-sample of the CWs onto the interday variance of daily use versus the average daily use, and the average daily baseline versus the average daily use. The colours used for each class are simply designed to aid visibility.

In Figure 4.2, we note the linear-like features of some CW clusters with high average daily use and high interday variance. These turn out to be customers with high use who are only active for typically two or three days per week.

In Figure 4.3, we show lines that indicate the trends expected if the average daily baseline use was 100%, 50%, and 5% of the average daily use. (Recall the

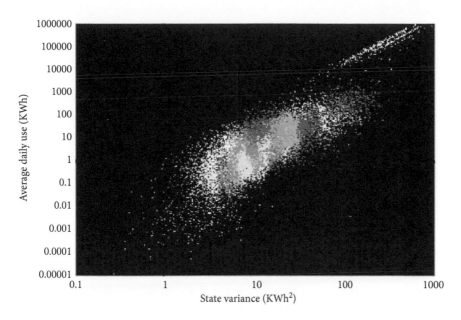

Fig 4.2 State variance (KWh²) versus average daily use (KWh).

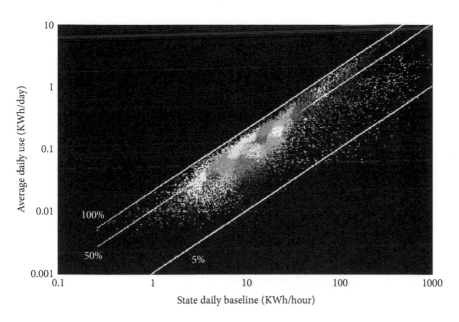

Fig 4.3 State daily baseline (KWh per hour) versus average daily use (KWh per day).

average daily baseline is a per thirty minutes value.) This shows how habitual is the demand exhibited within some of the CWs. At 100% every day is exactly the same, and equal to baseline demand: possibly the old, sick, or unemployed.

In work related to this analysis, we contrasted the behaviour-based segmentation with other segmentations available, for example that based on inferences of socio-economic class from postcodes (which are often used to target customers in acquisition programmes). There was poor correlation between the two (beyond a general trend) indicating that postcode is not a good predictor of demand and that customer to customer behavioural variation is really important. You are not like your neighbours!

4.5 Behaviour-Based Credit Assessment

There are two billion people in the world living within emerging economies who own a mobile phone but do not have a bank account. The first step to financial inclusion for such people is via a double 'leap frog'. They can become banked through mobile phones with simple cash-in-cash-out accounts attached to the SIM identity of their phones, enabling them to both send and receive money locally and internationally. Most users have prepay mobile phones, so converting their account balances to cash or paying-in may take place via the same network of outlets that vend the call credit. It is a double 'leap frog' because such nations never had sufficient fixed infrastructure, either for banking (banks could not deal with such numbers of active accounts) or for fixed phone lines. Instead, they jump straight into peer-to-peer mobile technologies offered by the Mobile Network Operators (MNOs) and the m-wallet banking proposition. The first major attempt to grow m-wallets through phones took place in Kenya with M-PESA growing to handle twice the number of mobile banking accounts than the number of real bank accounts, nationally, within just two years.

The second major step to financial inclusion is for such account holders to access some credit. Whether it is third parties (conventional banks, etc) that are lending to the MNO's m-wallet users, or the MNO itself (acting as a lending bank), this leads to a number of risks. First, that very little may be known about the account holders: they may be pre-pay and identified in a limited way, or even if known there may be little or nothing established about their lifestyle and real assets. Lending to such uncharacterized and unassessable borrowers cannot proceed along conventional credit scoring terms. Second, that the use of multiple SIM cards means that people may access credit and then simply vanish. Third, that the cost of lending may be quite high, since applicants may need to be contacted by call centres for example, yet many people who gain such a credit line (for a rainy day, or because they were asked) may never actually use it. This multiplies up the cost of acquisition for the remaining active borrowers.

Cignifi Inc, a Boston-based start-up company <http://www.cignifi.com>, has developed analytics technologies that resolve some of these problems by using possibly the only data which is not claimed and cannot be fabricated in order to credit score and qualify the potential borrowers. Their technology exploits raw call data records (CDRs) that establish the type of behaviour actually exhibited by the MNO's customers. The CDRs detail all calls (voice, SMS, data) made to, or from, the handset. Every CDR gives the time, duration, and other information about the calls such as location, the counter-party, and the call range (local, regional, national, international). Classically, the CDRs are used for billing, so are clearly available to the MNO, yet these are very big datasets (100s or 1,000s of CDRs per week, for millions of customers) so this can be compressed by summarizing the user's behaviour over suitable time periods, possibly through a very large number of attributes. The general approach is to employ a variant of the EM algorithm to segment customer behaviour over a relatively short time period (a few complete weeks) into a large number of states (classes).

Once a suitable finite mixture model is obtained and used to provide a dynamic segmentation of customer behaviour, this yields a class-based description of customers' recent behaviour over successive time periods. Then, based on some growing experience of real lending to some existing customers, there can be class-based scores defined for both credit worthiness and response to credit offerings.

All MNOs must create new value from their customers. The value of voice calls and SMS is decreasing, so m-commerce and m-leisure are natural candidates for growth. It has been shown that customers who get limited credit tend to use it first to convert themselves from pre-pay to post-pay, by funding calls with the credit. Then they become more loyal to the MNO and less inclined to swap SIMs. It does not stop at credit: there are many third party offerings that can be targeted to the customer segments most likely to be interested and respond reliably. This will undoubtedly become a core competence of MNOs in the future.

Together with Oi Telecom <http://www.oi.br.com> and the Inter-American Development Bank <http://www.iadb.org>, Cignifi illustrated the usefulness of dynamic behaviour-based credit worthiness, where customers qualify by their recent behaviour, with an application in northern Brazil that was published in a white paper in 2011 [26]. Although the fine details of the attribute extraction and model-processing remain proprietary, the principles enabling such applications should be clear. Mobile phone usage is not random: rather it tells us something about the lifestyle of the phone user. Meanwhile, that intelligence may be correlated to their credit worthiness and value as potential customers for new services and products. This case study showed that the discrimination achievable was effective, and that segments of high risk (four or five times the mean rate of loan delinquency after a number of months) and low/no risk (miniscule rates of delinquency) are readily identifiable from call-based behaviour alone.

4.6 Customer Missions in Consumer Goods Stores

'Customer missions' refer to the particular classes of shopping that a customer is undertaking during any particular transaction within a supermarket or some other multi-category store selling frequently-purchased consumer packaged goods. Each customer may carry out different transactions with a distinct 'mission in mind', at different times. If the retailer knows which missions are growing, which are most profitable, which are unique for them, which are hurt by key competitors' performance, which are driven by price or promotions, and so on, then they can invest some effort and resources strategically. Moreover, the national supermarket chains often have stores designed and ranged by different store formats (from local convenience, through superstores to hypermarkets, for example). These formats are designed to enable distinct types of missions to dominate (by concept at least). So the establishment of rigorous definitions for those missions that the retailer makes his strategic priorities is essential in justifying and measuring returns from large scale investments.

We might guess a set of missions (main family shopping, convenience shopping, distress/emergency, party-time, and so on), but how many are there, and how does week-part, day-part, the (category) content of the basket, and basket size contribute to definitions of missions? How can we define missions quantitatively?

A 'mission' is a collection of baskets that all share some similar properties. To define a set of missions, together describing all of the shopping baskets, we must introduce a set of attribution functions defined over a 'basket-property' space. Usually, when the missions are to be established by the observable data, the EM algorithm can be deployed, representing the totality of all shopping baskets as a finite mixture of baskets from distinct missions.

Each basket is to be represented by a set of measures or properties held in a vector in the form

$$\mathbf{x} = (x, y_1, y_2, \ldots, y_Q)^T.$$

Here, x is the natural logarithm of the total sales value (in pounds, say) of the total basket, and each y_r represents an integer denoting a categorical attribute. For example, y_1 might be an integer class number between one and seven, representing the day of the week that the basket was purchased, and y_2 might be an integer calls number between one and twenty-four, representing the hour of the day that the basket was purchased. For content, y_r might be an integer class number representing the number of items, j, purchased from a specific macro category in a geometric (logarithmic) scale $j \rightarrow g(j) = [\log_2(1+j)]^+$ (see Section 4.3).

There may be a number of macro categories: perhaps eight or ten, but not fifty (which would be cumbersome). Typical macro categories could be fresh fruit and veg; dairy; ambient food; beers wines and spirits; frozen foods; household goods; personal care; meat, fish, and poultry; pharmacy; and so on. Retailers have category hierarchies, so all items can be mapped into such macro categories. The same idea may also be used to count the number of items purchased that are on promotion, or that are own brand, and so on.

Hence the summary vector $\mathbf{x}$ is easily obtainable from the disaggregated basket data (essentially this is the till receipt listing time, day, location, quantity, and value for all items purchased). This data are called EPOS (electronic point of sale) data. Some retailers simply total up all till data by items (for supply chain purposes) almost immediately, so in that case they do not retain individual basket data, and thus they can never know anything about cross-purchasing or missions.

We can assume that there are $Q \geq 1$ such categorical (content counts, time, ...) variables. These, together with x, make up the vector, $\mathbf{x}$, which parameterizes 'basket property space'. A normal distribution for (log of basket value) x, and suitable multinomial distributions for each categorical variable, can be employed to define and discover a probability density distribution for each mission, preceding exactly as described in Section 4.2.

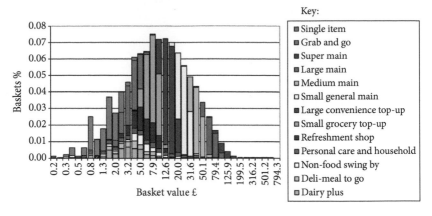

Fig 4.4 Mission distributions with respect to total basket value.

Here is the result of a simple EM-based missions analysis using baskets from a national retailer with hundreds of outlets (of various sizes and formats), analysing hundreds of thousands of baskets. Together, the shopping is split into thirteen distinct missions, see Figure 4.4. It clearly is not all about basket value. In fact, as you can see, total expenditure is a very poor discriminator of what the customer 'has in mind'.

There are, in fact, distinctive missions skewed towards categories, such as dairy, deli, personal care, and non-food. The larger, more valuable baskets (that is main shopping) are sub-divided into four natural missions, representing 18%, 21%, 20%, and 9% of total revenues, respectively. These missions are biased to take place at different times of the day and on distinct days of the week. They are also biased in terms of the amount of items purchased from different categories and sub-categories.

Figure 4.5 shows how the missions are distributed by outlet. In this case, the outlets are ordered in increasing order of the total fraction of main shopping (the top four missions). Clearly, instances (basket fraction) of main shopping (bottom) are only weakly related to the total outlet revenue (top). There are often some surprises here, where we find a large amount for 'grab and go' (snacking and lunching) at out-of-town superstores (easy parking for business travellers)! and some people having to carry out main shopping in convenience formats due to lack of transport or competitive alternatives. Stores may be grouped by format to look for aberrations, or else stores with distinct competitors (the nearest competitor chain for example) can be contrasted to

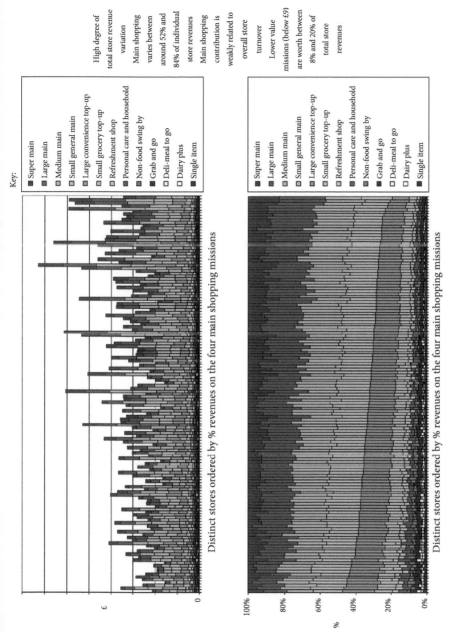

Key:

- Super main
- Large main
- Medium main
- Small general main
- Large convenience top-up
- Small grocery top-up
- Refreshment shop
- Personal care and household
- Non-food swing by
- Grab and go
- Deli-meal to go
- Dairy plus
- Single item

High degree of total store revenue variation

Main shopping varies between around 52% and 84% of individual store revenues

Main shopping contribution is weakly related to overall store turnover

Lower value missions (below £9) are worth between 8% and 20% of total store revenues

Distinct stores ordered by % revenues on the four main shopping missions

Distinct stores ordered by % revenues on the four main shopping missions

Fig 4.5 Mission distributions for different outlets (stores).

understand how the consumers perception of competitors ranges may drive losses of baskets.

Emerging in the early 2000s, and pioneered by Numbercraft Ltd [58] and others, this type of behavioural analytics is now very standard within retail. There may be many ways of defining missions, but a data-driven approach offered by variants of clustering algorithms is essential in sense-checking concepts offered by others and making sure that some class of customer behaviour has not just been missed, overlooked, or forgotten. If it exists as a behavioural cluster within the observable basket data, then it should be discoverable.

Of course missions work equally well online, but primarily it is a concept that works well for frequently purchased goods, so that missions are discovered through common behaviours over time. In fact it is the flexibility of missions defined via the EM algorithm and their ability to identify new (desirable and aberrant) classes of behaviour that is its key strength. So almost any customer-facing company with transaction/basket data available could consider how mission analysis, and data-driven missions in particular, can provide insights that drive growth.

4.7 Suggested work

From the website <http://www.oup.co.uk/companion/analytics>, download the test data sets CDR.txt and BASKET.txt together with their explanatory documents and associated tables. You can create customer segmentation and a missions segmentation, respectively.

Download the smartmeter data SMARTMETER.txt and the associated documents. Build a behaviour-based segmentation for such customers' demand profiles.

A Personal View

On Unsupervised Discrimination

Standard modules for clustering and EM algorithms are available in MATLAB, SAS, and other packages. Here, we have considered things from first principles so that readers could, if so inclined, employ their own bespoke distributions. In fact, the distributions offered in packages may be limited to real Gaussians, so multinomials and hierarchical models might be very useful. The warnings in the practicalities

---

section in this chapter cannot be stressed enough. One should be never naïvely apply EM-type methods. Just look at the data first.

Any method that segments customers' recent behaviour provides a dynamic read out when reapplied from time period to time period. This is invaluable, though it is often resisted by marketing departments who may wish to cling on to static segmentations. The value comes from the strategic insight, and the constant opportunities that behavioural change supplies. One might stand in the boardroom and ask 'How many distinct types of customer behaviour (not customer back-grounds) do you have?' and 'Do you react smartly when a customer's behaviour changes?'

On Dynamic Outputs

Whether we are segmenting behaviour, making inferences, or providing forecasts, it is essential to consider how our analytics will be deployed. A common mistake is to play down the dynamical elements within data. It is easy to think of very challenging and good applications of analytics that would produce little long-term insight and value. For example, suppose we are able to monitor the traffic on the streets of New York or pedestrian densities within London's roads and squares. If we produce a heat map (showing densities or similar quantities), then to be of value this must be able to change dynamically as new information comes in, so that we may respond to events and interventions. Otherwise one will just learn that there is a large crowd of people in Trafalgar Square and another in Piccadilly Circus. Of course there may be some surprises, but the point is that once seen, our users will not need to keep coming back to relearn these things. What they need is dynamic and preferably real-time insight: to compare now with then (yesterday, last week), or to look at the impact of key events. This sounds obvious, but it is a very common error. Behavioural changes drives value changes. In Chapter 7 we shall look at how segmentations made over relatively short time periods can become valuable, suggesting marketing tasks, estimating returns on investments (how many marketing departments really do this rather than just spending their budget?), and monitoring the response. Imagine the users of our analytics. We need to be sure they will wish to keep coming back and that our insights become an integrated part of their processes, changing every hour, every day.

On Probability and Functional Equations

Probability theory, and especially Bayesian probability theory, is often crowded out of our mathematics undergraduate courses, or else it is fleetingly ad-dressed through measure theory (analysis) or within stochastic dynamics

continued

On Probability and Functional Equations *Continued*

(random motions). This tends to avoid any statistical considerations in dealing directly with any actual data sets and plays down the subjective nature of probability. Do we build a house of cards, each with a high degree of rigour, yet ignore the subjective foundational caveats and assumptions that must be made?

The rules of conditional probability may be derived from functional equations. Long ago, functional equations were important within mathematics, as important as differential equations, but in most fields they lost out. Euler's functional equation for example, $f(x)f(y) = f(x + y)$ which has the solution $f(x) = \exp(\text{constant } y)$, is useful in many fields of mathematics. There is something elegant and foundational about functional equations: they crop up at exciting moments, such as within transitions to chaos [42]. Recursion and fractals are fundamental ideas and objects that are best expressed through functional equations. There has been a huge amount of interest in linear dilation equations, since these are satisfied by wavelets [118]; such functional equations are quite common.

Aczel's book [1] provides a historical and systematic treatment of the solution of functional equations, one of the oldest topics within mathematical analysis. In Kuczma, Choczewski and Ger (1990) [89] a further comprehensive overview and development is given.

Perhaps undergraduate courses would be the better for finer consideration of these topics. Let us claim these back!

CHAPTER 5

Multiple hypothesis testing over live data

5.1 Introduction to Multiple Hypotheses

In this chapter we shall consider a very general class of problem in which we receive successive observations and information and we must decide whether one of a number of distinct things is happening, possibly in real time as new data come in. We shall be employing the Bayesian framework that is introduced in the Appendix. For example, we may be observing customers on a website or players in an online casino, and we need to decide if they are behaving in an ignorant way (and need information or training), inconstantly (and need help), fraudulently (and need intercepting), or valuably (and need rewarding so as to build loyalty). Alternatively, we may be observing output (emails, blogs, messages) and wish to establish the mode of activity taking place or some aspect of the subject's identity (age, gender, etc).

In all of these problems we must use successive evidence to build support for each one of a number of mutually exclusive hypotheses, and then act when we need to draw a conclusion or when we have an opportunity to intervene, by adopting the hypothesis that is (at that moment) the most likely to be true.

Consider the problem of observing a sequence of events emanating from some unknown source. Suppose that the source may be in exactly one of N possible modes or, equivalently, may be of exactly one of N (≥ 2) possible types.

Mathematical Underpinnings of Analytics. First Edition. Peter Grindrod.
© Peter Grindrod 2015. Published in 2015 by Oxford University Press.

We wish to use all of our knowledge of similar past events, about the behaviour of such sources in the alternative modes and the newly observed events, so as to make a decision as to which mode the source is likely to be in.

5.2 Mathematical Framework

This type of problem requires us to test simultaneously and to complete the N hypotheses, denoted by H_k, meaning that the source is in mode k ($k = 1, \ldots, N$). After observing some events, we will have a current best (most likely) hypothesis and we may prefer one so much that we may take some decision consequent to it being correct. Let X denote our prior information of all events and all we know about the alternative modes. Let D denote the newly observed 'event' data. When D changes, our (posterior) estimates for the probability of each hypothesis, $P(H_k|D, X)$, will be updated via Bayes' Theorem, depending upon both the priors and the set of N models for the 'event' data, D, under the alternative hypotheses.

Just as in the Appendix, we let

$$O(H_k|X) = P(H_k|X)/(1 - P(H_k|X))$$

and

$$O(H_k|D, X) = P(H_k|D, X)/(1 - P(H_k|D, X))$$

denote the prior and posterior odds that H_k is true.

Then using (A.6), we have

$$P(H_k|D, X) = \frac{P(D|H_k, X)P(H_k|X)}{P(D|X)} = \frac{P(D|H_k, X)(1 - P(H_k|X))O(H_k|X)}{P(D|X)}$$

and

$$P(notH_k|D, X) = \sum_{j \neq k} P(H_j|D, X) = \sum_{j \neq k} \frac{P(D|H_j, X)P(H_j|X)}{P(D|X)}$$

and hence

$$O(H_k|D, X) = \frac{P(D|H_k, X)}{\sum_{j \neq k} P(D|H_j, X)\frac{P(H_j|X)}{(1 - P(H_k|X))}} O(H_k|X). \tag{5.1}$$

Compare this with (A.9) which was valid for $N = 2$ and is thus simpler. The terms $P(H_j|X)$ (appearing within both the prior odds and within the denominator of the updating term) are the priors. If $N = 2$, then notice that the updating term simplifies to a simple likelihood ratio: $P(D|H_k, X)/P(D|H_j, X)$ (where j is not k) as in (A.9), but this is not our general situation.

This often causes confusion. If we have only two hypotheses (A and $notA$), the odds are updated simply by multiplying by a Bayes factor which contains only model terms (the ratio of the probability of the data under the alternative hypotheses). For N greater than two, this simply does not happen, and to update the odds we must multiply by a factor that involves both the model terms and the priors. That is how it is.

Now consider the general set up: look at the updating terms in (5.1), that is, the ratios

$$\frac{P(D|H_k, X)}{P(D|not\ H_k, X)} = \frac{P(D|H_k, X)}{\sum_{j \neq k} P(D|H_j, X) \frac{P(H_j|X)}{(1-P(H_k|X))}}$$

which update the odds $O(H_k|D, X)$ in (5.1). In order to proceed, we simply need a set of models for the 'event' data, D, under the alternative hypotheses, H_j, for $j = 1, \ldots, N$. Then successive independent observations can allow us to successively update the odds on all hypotheses.

In applications, we often want to describe incoming events in terms of performance measures, called extracted 'features'. As we observe a new event, we summarize it in terms of a number of features, which will be assumed to be bounded real numbers. Then under each separate hypothesis we will require a model: a probability distribution defined for the features-vector values (the 'support set' in feature-space) under each hypothesis. A very good idea, discussed in Section 5.3, is to convert the density distribution over the support set, a cube in real n-space, into a multinomial distribution, by partitioning that cube appropriately, so as to approximate the density function for a continuous variable (the features) with a discrete multinomial distribution over a partition of feature space. This is of course particularly useful when there are not a lot of observations present.

5.3 Multinomial Approximation of a Continuous Density

In this section we consider *some routes by which we might condition the models* based on observations. Suppose that the observation of an event is represented by some vector, $\mathbf{z}$, of measurable quantities, performance measures, or classifications. We will call $\mathbf{z}$ the feature vector, and say that it lives in a feature space.

Suppose that there are J features, so $\mathbf{z}$ is of length J. Here we will assume that these measurements are real, or at least integers. The extension to other types of variable (categorical for example) is obvious.

Now suppose that the feature space is partitioned into exactly m subsets, here called classes, $C_1, \ldots, C_m$. In reality, under hypothesis H_k, any $\mathbf{z}$ that is observed is drawn from some (unknown, yet sampled) density distribution, $f_k(\mathbf{z}|X)$, that is defined over the feature space.

For our purposes, we may use a multinomial model, induced by the partition, $\{C_r\}$, so as to represent a simplified distribution for each of the $f_k(\mathbf{z}|X)$, the feature density distributions under each of the $k = 1, \ldots, N$ hypotheses.

Set

$$P_{k,r} = P(\mathbf{z} \in C_r|H_k, X) = \int_{C_r} f_k(\mathbf{z})d\mathbf{z}$$

to denote the probability that a random $\mathbf{z}$ observed under H_k lies in C_r assuming all our prior experience X.

In practice we may not know f_k or the $P_{k,r}$ very accurately. So we have to estimate the $P_{k,r}$ based on some sets of observations. We may use Laplace's law of succession to do so, based on a sample $\tilde{\mathbf{z}}$ observed under H_k. This is fully discussed in the Appendix. We have the estimate for $P_{k,r}$ given by

$$the \hat{P}_{k,r} = \frac{\text{number of } \tilde{\mathbf{z}} \in C_{k,r} + 1}{\text{total number of } \tilde{\mathbf{z}} + m}.$$

This is a low-resolution approximation to the full density function. Clearly, its calibration is not sensitive to any errors in the $\tilde{\mathbf{z}}$s that do not perturb them across the partition.

How can we generate suitable partitions? The easiest way is to generate m **centroids** (m feature vectors within the feature space) and partition feature space by sending all possible feature vectors to the nearest centroid.

Finally returning to the updating: from (5.1) we have

$$O(D|H_k, X) = \frac{P(D|H_k, X)}{\sum_{j \neq k} P(D|H_j, X) \frac{P(H_j|X)}{(1 - P(H_k|X))}} O(H_k|X).$$

If the incoming data, D, are simply the observation that $\mathbf{z} \in C_r$ and $N = 2$, then this simplifies (just as we observed earlier) and then taking logs we have a simple updating rule, say:

$$\log O(D|H_2, X) = \log \frac{\hat{P}_{2,r}}{\hat{P}_{1,r}} + \log O(H_2|X). \quad (5.2)$$

Exercise 5.1

Derive (5.2).

5.4 Example: Beatles or Stones?

In this section, we want to present a monitoring application that will read the lyrics of a song and constantly update the estimate as to whether the song was written by either the Beatles or the Rolling Stones. This problem is clearly analogous to a number of monitoring and decision problems in business, in particular where we can monitor performance in real time. It has rather soft data, so we have to make some quantitative measurements under each hypothesis (that the lyrics were written by either the Beatles or the Stones).

First, we need some calibration data and we will deliberately keep things very simple. We selected five songs from each band (mid-sixties and early-seventies tracks, so only vintage Stones). Each song was chopped up into distinct one-hundred-character length, 'bite sized', pieces (including spaces and punctuation characters). These pieces will be called 'bites'.

The lyrics obtained yielded thirty-five Beatles' bites, forty-two Stones' bites, and ten bites from another 'mystery song' that will represent our input data, and where we want to attribute authorship. Then for each 'bite', we made counts of the number of vowels and the number of punctuation marks within them.

We normalized the counts so that all values were mapped onto the interval zero to one. So each bite became a $\mathbf{z}$ on the unit square. For the vowel count we defined

$$z_{vowel} = \frac{\text{number of vowels}}{14} - 3.$$

For the punctuation mark count we defined

$$z_{punc} = \frac{\text{number of punctuation chars}}{7}.$$

Hence for each bite we set $\mathbf{z} = (z_{vowel}, z_{punc})$.

Next, we generated five centroids as random vectors within the unit square. For both the calibration bites (under both hypotheses) and the bites from a mystery song, we mapped all $\mathbf{z}$s to the class corresponding to the centroid that was closest using the usual Euclidean (unweighted sum of squares) distance measure. In this way the $\mathbf{z}$s were partitioned into five classes.

The use of centroids (and possibly weights for the Euclidean metric) is a highly useful way of generating partitions.

The centroids used were as follows:

$$(0.542, 0.473)^T, (0.47, 0.524)^T, (0.521, 0.468)^T, (0.547, 0.485)^T, (0.467, 0.463)^T.$$

For the calibration sets of the Beatles and Stones data, we used Laplace's law of succession to yield the following estimates of the multinomial probabilities of $\mathbf{z}$ being within each cell, conditional of each hypothesis (designating authorship):

$$(P_{Beatles,1}, \ldots, P_{Beatles,5}) = (0.125, 0.075, 0.150, 0.025, 0.625)$$
$$(P_{Stones,1}, \ldots, P_{Stones,5}) = (0.234, 0.128, 0.128, 0.0426, 0.468).$$

These multinomials are the two models: $P(\mathbf{z}|H_{Beatles}, X)$ and $P(\mathbf{z}|H_{Stones}, X)$.

Next we selected our priors $P(H_{Beatles}|X) = P(H_{Stones}|X) = 1/2$ to represent our starting point of indifference.

So the prior odds on the Beatles $O(H_{Beatles}|X) = P(H_{Beatles}|X)/(1 - P(H_{Beatles}|X))$ are one.

Successively, the ten bites from the mystery song were in classes 5, 3, 3, 5, 5, 5, 5, 5, 3, and 5. After each bite we recomputed the odds using (5.1) (each time multiplying the previous odds by the updating term). We obtained

$$O(H_{Beatles}|Bites, X) = 1.34, 1.57, 1.84, 2.46, 3.28, 4.39, 5.86, 7.82, 9.19, 12.27,$$

respectively.

After analysing the lyrics, the method asserts that it is twelve times more likely to be a Beatles song than not.

The mystery song was *Strawberry Fields*.

This was a very simple example, and had some of the bites from the mystery song appeared within classes one or four, then the odds would have been reduced at the corresponding updating step. However, it is important that we draw confidence from this example. It was simple and successful, and at any point we could have given our best estimate of attributed authorship, so this method can be used in real-time monitoring situations.

In practice, we should next make some attempts to optimize the model. The method of log Bayes' factors would be a good one here (see Section 3.8). In such a method, we estimate the evidential value of a bite appearing in each class. Then we take a population average of such values, so that we get a typical evidence value from a single bite. Clearly we want to choose a partition where this is large, and hence the model is likely to discriminate with as little data as possible.

Note that this example was really easy to compute because we had only two hypotheses competing, and recall that the updating formula is particularly sweet in that case, since the updating multiplier to be applied to the odds is independent of the priors.

Using the definitions supplied above for z, based on one hundred consecutive characters from the lyrics (including spaces and punctuation), the centroids, and the probabilities from the calibration sets (again all given above), you can now run your own tests on your favourite Beatles or Stones songs. Using the lyrics to *Angie*, as a second 'mystery' example, then successively the nine bites from this mystery song are in classes 1, 1, 4, 3, 1, 1, 1, 1, and 1. After each bite we obtained

$$O(H_{Beatles}|Bites, X) = 0.53, 0.28, 0.17, 0.20, 0.10, 0.06, 0.03, 0.02, 0.01.$$

Notice the fourth bite is in class three, and is the only time the odds lean away from the Stones.

5.5 Identifying Spinners on Mobile Phone Networks

In our next example, which is a matching or identification problem, we do not use a calibration data set in the way we have in Section 5.4. Instead, we shall define the models by directly querying the data generated in the past under each of a large number of hypotheses, using a simple model to estimate the probability of generating the set of current observations, given this history in each case.

Spinners are mobile phone users who change their Mobile Network Operator (MNO) at high frequency to take advantage of new promotions and deals (they are promotion junkies). Spinners often change their SIM cards and are difficult to detect for MNOs. The only information the MNOs are able to use is what they observe within the calling patterns of their customers (who they call and where they place calls). The MNOs gather the logs of all calls and text messages of all their users, and hence they can know who calls, and is called by, each of their users. These logs are called 'call data records', or CDRs for short. CDRs contain the time of incoming and outgoing voice calls and texts as well as the location of the users at the time of the call, allowing them to estimate their preferred locations. Can the MNO decide if a new user is in fact a spinner?

This is a Bayesian multiple hypothesis question.

Let us assume the MNO has a (possibly rather large) set of known or assumed lapsed users, say $\{U_1, \ldots, U_N\}$ for N large, and they have historical data from their CDRs. Now suppose that we have some new data generated by an unknown new user, U say. Think of this as a number of new observations under the hypothesis, H_k say, that the new user $U = U_k$ ($k = 1, \ldots, N$). These are mutually exclusive.

The historical data may be used to create a 'model' of various observations, the evidence E, under H_k. That is the probability $P(E|H_k, X)$ of observing evidence E given H_k. This must be well defined for various types of evidence E.

Obviously we shall need a model for all $k = 1, \ldots, N$, and each must be efficient.

We have priors $P(H_k|X)$ which could be uniform, unless we have any other information (in X) such as region or location of users. Suppose all U_ks and U are from one location (or segment, or even that they are all customers buying SIMs and call credit from a certain group of shops/outlets). Then set $P(H_k|X) = 1/N$, so this prior is uniform and plays no further role.

Then Bayes' rule gives us the non-normalized posteriors:

$$P(H_k|E, X) = P(E|H_k, X).$$

Suppose, for example, that we observe new user U call a person A either because he/she knows them and has called them in the past or because he calls them for some new (possibly random) reason. Then for each lapsed user, U_k, we can quickly see how many times that they have called A within their available history. If U_k called A an average number of times, $r_k(A)$, within some given time period (such as a week, so that $r_k(A) = 1$ means once within the week, $r_k(A) = 7$ means every day), perhaps $r_k(A)$ could be an average call frequency (over a few weeks).

Then we might employ a Poisson distribution, assuming H_k (that $U = U_k$), as a model for the (integer) possible number of times, r_A say, that U was observed to call A within such a time period. Calibrating the Poisson distribution from U_k's history, we have $E = $ 'U called A exactly r_A times' and

$$P(E|H_k) = r_k(A)_A^r \frac{e^{-r_k(A)}}{r_A!}.$$

So under H_k, the expected number of calls that $U = U_k$ would make to A is $r_k(A)$, and the variance is $r_k(A)$ also. Note that $r_k(A)$ is very inexpensive to query from the given historical data. So we can assume it is known for every person A that U actually called. Since N is generally very large, we need to avoid doing sophisticated things with that historical data, such as creating a model for all possible events. Instead, we only query the information we need from each U_k corresponding to the small amount of behaviour that is observed from U.

Yet we do not quite wish to employ this immediately since we may historically find that $r_k(A) = 0$ (that is U called a person A that we never saw U_k call historically). Obviously, there must be some small chance that U_k might call somebody who he had never called before, so we shall adjust this estimate by a small amount and define $r_k(A) = \delta > 0$ for some δ fixed that $U = U_k$ now calls a party A who they did not call previously. We shall never estimate $r_k(A) = 0$. We might be more subtle than this, but let us assume $\delta = 10^{-3}$ or so for now. Then if E says U calls A exactly r_A times, and $r_k(A) = \delta$, then

$$P(E|H_k) = \delta^{r_A} \frac{e^{-\delta}}{r_A!} \sim \delta^{r_A} \frac{(1-\delta)}{r_A!}.$$

Now we make E much bigger, and assume that we observe many of U's contacts, a set of As of size M, and that the numbers of calls made to each distinct counter party A are independent, so that we may multiply probabilities. Suppose we observe each person A who U called a number $r_A \geq \delta$ times in a given time period (or a frequency r_A). Then for all $k = 1, \dots N$, taking logs, we have

$$\log P(H_k|E) = \sum_A \log \left(r_k(A)^{r_A} \frac{e^{-r_k(A)}}{r_A!} \right) = \sum_A r_A \log r_k(A) - r_k(A) - \log(r_A!).$$

In fact the last term, $\sum_A \log(r_A!)$, is the same for all k so in practice we may happily ignore it, since we only need to know the relative values of the $P(H_k|E)$, and any constant term such as this is a common normalization.

At any point (as we observe more and more calls from U and M increases) we could halt, and choose the modal H_k that gets the largest log-likelihood as the identity of a possible spinner, or else we may remain passive unless we exceed some threshold, dependent on the number of counter-parties considered, M. For example, we should divide $\log P(H_k|E)$ by M so that we are actually measuring and maximizing the Mth root of product of the M independent factors within $P(H_k|E)$. This trick would allow M to vary as U varies, or the addition of new counter-parties, should we need to set such a threshold-stopping rule.

Of course if U_k has a new friend A whom he/she calls a lot now (when identified as U) but that he/she did not call at all before (when identified as U_k), then the contribution to $\log P(H_k|E)$ from that particular A will be very negative indeed (since $r_k = \delta$, small):

$$r_A \log \delta - \delta - \log(r_A!).$$

So obviously spinners who change their best friends will be harder to identify.

On the other hand, if $r_A = r_k(A)$ then the contribution to $\log P(H_k|E)$ from that A will be less negative: we have

$$r_A \log r_A - r_A - \log(r_A!).$$

This last is interesting as it reminds us of Stirling's formula, a lovely result that gives us an approximation to $r!$, for $r \in \mathbb{Z}$ large:

$$r! = r^r e^{-r} \sqrt{2\pi r}(1 + O(1/r)).$$

Equivalently

$$\log(r!) = r \log r - r + \frac{1}{2} \log 2\pi r) + O(1/r).$$

It follows that if $r_A = r_k(A)$ and is large (that is, both U and U_k called A many, many times), then we will have a corresponding contribution to $\log P(H_k|E)$ given by

$$r_A \log r_A - r_A - \log(r_A!) = -\frac{1}{2} \log 2\pi r_A + O(1/r_A).$$

This decays very slowly as r_A becomes large.

Exercise 5.2

What about near misses? Suppose that we have $r_k(A) = r_A \pm 1$. Then the contribution to $\log P(H_k|E)$ is still

$$r_A \log r_A - r_A - \log(r_A!) = -\frac{1}{2} \log 2\pi r_A + O(1/r_A).$$

Exercise 5.3

What about other misses? Suppose that we have $r_k(A) = r_A \pm z$. What is the contribution to $\log P(H_k|E)$ for large r_A and fixed z? (The relative miss gets smaller and smaller.)

5.6 Logistic and Polytomous Regression

In this section, we consider some classical ways in which models might be calibrated for each of a number N of competing hypotheses. The simplicity, the limitation, and also the acceptability of these models come from their linear *scorecard* nature. We shall assume that we are given a calibration data set of observations (a set of attributes) labelled by the relevant hypothesis that has generated them. Of course we need samples under each separate hypothesis. Effectively each hypothesis has a score that is a linear functional that is to be determined from the data over the observed attribute space, which is some domain within $\mathbb{R}^p$. Thus for the kth hypothesis, H_k, the scores appended to observations are simple functionals of the form $s = \beta_k^T \mathbf{x}$ for some $\beta_k \in \mathbb{R}^n$. Then these scores are mapped onto probabilities monotonically. Bearing this in mind, it is a very good idea to eyeball the data beforehand, by scattering the observations by class with respect to any two of the given real or categorical attributes. If the transition between the class is not generally directional (for example, if one class roughly surrounds another within such a 2D projection, or they have alternating regions of high density in some direction), then one really needs to select some other coordinates (mappings from the original attributes) that will make it so. If classes just look similar or confused, then that does not matter so much; those attributes might be worthless in terms of discrimination between some pair. This, of course, is easier said than done in some cases, but the point is that if the classes alternate their local modes and there is no clear directional separation, then these *generalized linear methods* cannot work well at all.

Having said that, when used properly these methods can provide good classification, and bearing in mind the workings of the method, which is fully presented below, one may be able to increase the efficacy by a careful choice of transformation of the basic variables. This will work well when the transition between any single hypothesis and the rest is at best close to a directional linear functional, and at worst confused.

These are amongst the most well-used methods within supervised discrimination (or classification as some call it). Such models attach probabilities to the proposition that some evidence, E, observed has emanated from just one of a number of types of source, or class, or hypothesis H_k $k = 0, 2, \ldots, N - 1$. The mathematical derivation of these models, $P(E|H_k)$, is often absent from accounts in textbooks though. Here we start with $N = 2$.

Logistic Regression

Let $\mathbf{x} = (1, x_2, \ldots, x_p) \in \mathbb{R}^p$ be a vector of observed attributes. If we observe categorical attribute a with m possible categories then we may unpack that information into $m - 1$ binary variables, $\mathbf{w} \in \{0, 1\}^{m-1}$, where $w_j = 1$ if, and only if, a is in the jth category ($\mathbf{w} = (0, \ldots, 0)^T$ if a is in the mth category). Then we shall assume that $\mathbf{w}$ is held within $\mathbf{x}$.

We shall write $\mathbf{x} \sim H_k$ to mean $\mathbf{x}$ was generated by hypothesis H_k.

Next, suppose we are given n observations $\{\mathbf{x}_i\}$ together with a given set of binary indicator variable $\{y_i\}$ such that $y_i = 0$ iff $\mathbf{x}_i \sim H_0$ and $y_i = 1$ iff $\mathbf{x}_i \sim H_1$.

For any $\beta \in \mathbb{R}^p$, the product $s = \beta^T \mathbf{x}$ denotes a *score*. We shall search for a suitable choice of β for which the probability $P(H_k|\mathbf{x}, \beta, X)$ is given by a function f of the score $s = \beta^T \mathbf{x}$. What is f? The scores may range over $\mathbb{R}$, so we shall select an f that maps $\mathbb{R}$ onto $[0,1]$ so it can represent a probability. A range of distinct fs might be useful, but we shall choose

$$f(s) = \frac{e^s}{1 + e^s}, \tag{5.3}$$

since it will make the algebra elegant. Then the log of the odds,

$$O(H_1|\mathbf{x}\beta, X) = \frac{P(H_1|\mathbf{x}, \beta, X)}{(1 - P(H_1|\mathbf{x}, \beta, X))} = \frac{P(H_1|\mathbf{x}, \beta, X)}{P(H_0|\mathbf{x}, \beta, X)},$$

is given by

$$\log \frac{f(s)}{1 - f(s)} = s = \beta^T \mathbf{x}.$$

Thus the linear functional of $\mathbf{x}$ is equivalent to the log odds that $\mathbf{x} \sim H_1$. This choice of f in (5.3) is called the logit function.

Notice that directly from (5.3), f satisfies the ODE

$$\frac{df}{ds} = f(1 - f). \tag{5.4}$$

Other sensible choices for f might be solutions of

$$\frac{df}{ds} = f(1 - f)R(f)$$

for some strictly positive function R on $[0,1]$. This is discussed in McCullagh and Nelder (1989) [102].

Assume that the observations are independent. Then the combined probability of observing that data set is called the likelihood, $\mathcal{L}$:

$$\mathcal{L} = \prod_{i=1}^{n} P(H_1 | \mathbf{x}_i, \beta, X)^{y_i} (1 - P(H_1 | \mathbf{x}_i, \beta, X))^{1-y_i}.$$

Notice for each i only one term appears on the right. We have

$$\mathcal{L} = \prod_{i=1}^{n} f(\beta^T \mathbf{x}_i)^{y_i} (1 - f(\beta^T \mathbf{x}_i))^{1-y_i}.$$

Taking logs:

$$\log \mathcal{L} = \sum_{i=1}^{n} y_i \log f(\beta^T \mathbf{x}_i) + (1 - y_i) \log(1 - f(\beta^T \mathbf{x}_i)).$$

Rearranging we obtain

$$\log \mathcal{L} = \sum_{i=1}^{n} y_i \beta^T \mathbf{x} + \log(1 - f(\beta^T \mathbf{x}_i)). \tag{5.5}$$

Now we must choose β so as to maximize this.

Using (5.4) to deal with the f terms, we may obtain

$$\frac{\partial}{\partial \beta_j} \log \mathcal{L} = \sum_{i=1}^{n} \mathbf{x}_{i,j}(y_i - f(\beta^T \mathbf{x})).$$

(5.6)

Here $\mathbf{x}_{i,j}$ denotes the jth element of $\mathbf{x}_i$.

Similarly

$$\frac{\partial^2}{\partial \beta_j \, \partial \beta_{j'}} \log \mathcal{L} = \sum_{i=1}^{n} -\mathbf{x}_{i,j}\mathbf{x}_{i,j'}f(\beta^T \mathbf{x})(1 - f(\beta^T \mathbf{x})).$$

(5.7)

We set $\log \mathcal{L} = 0$ in (5.6) for the *normal* equations, to be satisfied at a maximum. In practice we cannot solve these equations by hand, but the right-hand sides of (5.6) and (5.7) yield the gradient and the Hessian of the log likelihood, which is to be maximized as a function of $\beta \in \mathbb{R}^p$. So standard gradient methods available in numerical packages will locate β and more besides (for example, see Press et al (2007) [111]).

From this analysis, we can see that this method works well if we have data, or if we can transform the data so that the two classes are (relatively) separable by a linear functional. As mentioned at the start of this section, if this is not the case, for example one set of observations in $\mathbf{x}$ space surrounds the other, then transforming the observed variables into new coordinates may well help a lot. Of course, this assumes that the attributes may be represented by reals and thus embedded in a Euclidean space, yet recall logistic regression also is often applied with (unpacked) categorical variables, as indicated above.

There is a methodology [31, 111] that can search for suitable transformations in the case that there are two classes corresponding to two competing hypotheses (just as for logistic regression) and the observations may be all treated as real vectors. *Support vector machines* (SVMs) seek to relocate those observations by mapping them from the original space into another possibly higher dimensional Euclidean space. There is a mapping $M : \mathbf{x} \in \mathbb{R}^p \rightarrow \mathbf{z} \in \mathbb{R}^q$, where $q \geq p$. In doing so, it is designed to make scalar 'dot' product within $\mathbf{z}$-space easily calculable in terms of a *kernel* function, defined for the corresponding original vectors in $\mathbf{x}$-space: we have

$$M(\mathbf{x}_a)^T M(\mathbf{x}_b) = K(\mathbf{x}_a, \mathbf{x}_b)$$

for all $\mathbf{x}_a$ and $\mathbf{x}_b$. This makes linear algebra in $\mathbf{z}$-space efficient and relatively uncomplicated, and it is elegant. Hyperspaces in $\mathbf{z}$-space are thus given by level sets of the form $\{\mathbf{x}|K(\mathbf{x}, \mathbf{x}^*) = \text{constant}\}$ for some $\mathbf{x}^*$ fixed, and a given constant. For example, a common choice is to use radial basis functions as a kernel:

$$K(\mathbf{x}_a, \mathbf{x}_b) = \exp(-||\mathbf{x}_a - \mathbf{x}_b||^2/2\sigma^2).$$

There is a very large literature on this approach and its generalizations, in establishing a suitable K efficiently so as to separate the two classes of points (and exploiting this in various ways). SVMs belong to a family of generalized linear classifiers and they maximize the linear separation of two data sets, usually called the 'geometric margin', thus they are sometimes called maximum margin classifiers. The performance of an SVM depends on the selection of the class of kernel and its parameterization. The parameters of such a resolved model are sometimes difficult to interpret as necessarily they involve nonlinear combinations of some, or all, of the original variables. Interpretation is much easier though for logistic regression, since those models are explicable in terms of the score s and the vector β which we can refer to as a scorecard. Given the variation observed across all observations within each coordinate of $\mathbf{x}$, we can say which coordinates are most valuable, or the most predictive.

There is a further related issue for models used as classifiers such as SVMs, logistic, and generalized regressions: it is that of transparency. If we apply a model, then users, stakeholders, and sometimes the individuals who are themselves being classified will seek to know (and may have a legal right to be told) why they have been scored and classified thus. If the method is complex, non-linear, and requires the calculation of hidden intermediates, then it simply is not transparent. A failure to make the method intuitive or probable may result in it being mistrusted. When the 'computer says no!', what right of appeal or pathology do individuals have?

In passing, we note that other 'black-box' classification methods, such as those based on artificial neural networks, have a similar Achilles heel. Though popular in machine learning, where perhaps the performance trumps the lack of transparency, such methods may contain hidden interpolations and extrapolations, and they may be data hungry. It would be fine to apply such approaches in providing applications such as face recognition, or within applications used by all of us on platforms such as Facebook or Google, where the public users accept that they will occasionally perform poorly but willingly use them in a non-critical spirit. Yet it is difficult or impossible to apply such classifications in fields subject to regulatory controls and standards or even legal rights of investigation or appeal, such as health and safety, clinical, financial, or legal situations.

Scores and evidence in sequential updating of the log odds

Now let us assume that we have a logistic model calibrated and return to the context of the beginning of this chapter. There we receive a set of data, D, usually a live sequence of observations $\mathbf{z}_i$ for $i = 1, 2, \ldots$, all generated by the same source for which either H_1 or H_0 is true. Assuming independence of the successive observation we have

$$\log O(H_1|\{\mathbf{z}_j|j = 1, \ldots, i\}, \beta, Z) = \beta^T \mathbf{z}_i + \log O(H_1|\{\mathbf{z}_j|j = 1, \ldots, i - 1\}, \beta, X).$$

Thus for the sequence to date we have

$$\log O(H_1|\{\mathbf{z}_j|j = 1, \ldots, i\}, \beta, Z) = \sum_{j=1}^{i} \beta^T \mathbf{z}_j + \log O(H_1|\beta, X).$$

The term $\log O(H_1|\beta, X)$ is simply the prior—our estimate before we have seen any data.

For each observation, the corresponding score is equal to the change in the log odds. This change is sometimes called the evidence [83] and updating the log odds is thus very simple. This last equation is analogous to equation (5.2) that we derived for $N = 2$ when we were using multinomial models rather than logistic models.

Polytomous Regression

This is more of the same concept, yet the discrimination is between $N > 2$ classes, and thus more involved. If readers can write a code to implement logistic regression, then they should be able to implement this too. We include it here for completeness, and to satisfy the curious.

Here, as before, let $\mathbf{x} = (1, x_2, \ldots, x_p) \in \mathbb{R}^p$ be a vector of observed attributes. Again, we are given N such observations $\{\mathbf{x}_i\}$ and $\mathbf{x}_{i,j}$ denotes the jth element of $\mathbf{x}_i$. Now we are also given a set of binary indicator vectors $\{\mathbf{y}_i\} \in \{0, 1\}^N$, where here we shall index the elements of $\mathbf{y}$ by $k = 0, 1, \ldots, N - 1$, and we have $\mathbf{y}_{i,k} = 1$ iff $\mathbf{x}_i \sim H_k$ for $k = 0, \ldots, N - 1$, and zero otherwise. So $\mathbf{y}$ points to $\mathbf{x}$'s class.

Let us have N vectors $\beta_k \in \mathbb{R}^p$: where $\beta_0 = (0, 0, \ldots, 0)^T$, while the others are still to be determined from the calibration data.

Then we write

$$P(H_k | \mathbf{x}, X) = \frac{e^{\beta_k^T \mathbf{x}}}{\sum_{k'=0}^{K} e^{\beta_{k'}^T \mathbf{x}}} \equiv \pi_k(\mathbf{x}).$$

It is clear that these sum to unity. Moreover

$$\log \left(\frac{P(H_k | \mathbf{x}, X)}{P(H_0 | \mathbf{x}, X)} \right) = \beta_k^T \mathbf{x}.$$

Notice too that after some rearrangement, for each i:

$$\frac{\partial \pi_k(\mathbf{x}_i)}{\partial \beta_{k,j}} = \mathbf{x}_{i,j} \pi_k(\mathbf{x}_i)(1 - \pi_k(\mathbf{x}_i)) \tag{5.8}$$

where $\beta_{k,j}$ is the jth component of β_k. Also

$$\frac{\partial \pi_k(\mathbf{x}_i)}{\partial \beta_{k',j}} = -\mathbf{x}_{i,j} \pi_k(\mathbf{x}_i) \pi_{k'}(\mathbf{x}_i). \tag{5.9}$$

Now we are ready to form the likelihood:

$$\mathcal{L} = \prod_{i=1}^{n} \prod_{k=0}^{N-1} \pi_k(\mathbf{x}_i)^{\mathbf{y}_{i,k}}.$$

Then after some rearrangement we have

$$\log \mathcal{L} = \sum_{i=1}^{n} \left(\sum_{k=0}^{N-1} \mathbf{y}_{i,k} \beta_k^T \right) - \log \left(\sum_{k'=0}^{N-1} e^{\beta_{k'}^T \mathbf{x}_i} \right).$$

We can optimize this last with respect to $\{\beta_k | k = 1, \ldots, N-1\}$. The normal equations are obtained from the zero of the gradient:

$$\frac{\partial \log \mathcal{L}}{\partial \beta_{k,j}} = \sum_{i=1}^{n} \mathbf{x}_{i,j}(\mathbf{y}_{i,k} - \pi_k(\mathbf{x}_i)),$$

where $j = 1, \ldots, p$ and $k = 1, \ldots, N-1$.

Using (5.8) and (5.9) we can construct the Hessian from

$$\frac{\partial^2 \log \mathcal{L}}{\partial \beta_{k,j} \partial \beta_{k,j'}} = -\sum_{i=1}^{n} \pi_k(\mathbf{x}_i)(1 - \pi_k(\mathbf{x}_i))\mathbf{x}_{i,j}\mathbf{x}_{i,j'},$$

and

$$\frac{\partial^2 \log \mathcal{L}}{\partial \beta_{k,j} \partial \beta_{k',j'}} = \sum_{i=1}^{n} \pi_k(\mathbf{x}_i)\pi_{k'}(\mathbf{x}_i)\mathbf{x}_{i,j}\mathbf{x}_{i,j'}.$$

Hence, we may determine $\{\beta_k | k = 1, \ldots, N-1\}$ using a gradient search [111].

Many packages provide polytomous regression options, but it may be as well to (i) know what is going on inside; and/or (ii) have an option to simplify things in any given situation. One can see how critical the linear functional $\beta_k^T \mathbf{x}$ is for each class.

A Personal View

On Classifiers

In 2006, Hand [69] wrote 'A great many tools have been developed for supervised classification, ranging from early methods such as linear discriminant analysis through to modern developments such as neural networks and support vector machines. A large number of comparative studies have been conducted in attempts to establish the relative superiority of these methods...these comparisons often fail to take into account important aspects of real problems, so that the apparent superiority of more sophisticated methods may be something of an illusion. In particular, simple methods typically yield performance almost as good as more sophisticated methods, to the extent that the difference in performance may be swamped by other sources of uncertainty that generally are not considered in the classical supervised classification paradigm.'

There are two types of application for discrimination methods: those tasks that require a fast decision which we accept may make some errors, and those tasks where we need rigorous and defensible decisions. Our brains are really good at fast decisions (that is what heuristics can do [84]), and there are applications where we can accept occasional failures, such as face recognition and tagging within photographs on social media sites. More care must be taken for rigorous decisions where there is a high cost attributed to errors, and possibly a need for some transparency. In particular, if there are legal or regulatory issues, or rights to appeal decisions, or accountants of the board, then one cannot accept either occasional errors in performance or any lack of transparency.

Of course generalized linear models, including logistic and polytomous regression, have an appeal of simplicity and transparency. The analyst is in control of what is going on, and everything should be as simple as possible, but not simpler.

Personally, I greatly dislike artificial neural networks (ANNs). To me these extol a simplistic brute-force methodology, their workings are opaque, and they seem to require a very large amount of appropriate training. There are rather few commercially successful applications. I stand with Dewdney [35], '[Neural networks'] powers of computation are so limited that I am surprised anyone takes them seriously as a general problem-solving tool.' Worse, they are simply too hungry for computer memory and processing time.

What I object to most of all, though, is the appropriation of the clothes (and the language) of human neural cognitive processing. Your brain, merely 1.4 kg or so, relies on cell-to-cell interactions, each with nonlinear dynamics and significant transmission delays, set within an architecture. It is occasionally logically and arithmetically fallible, but it generally works well, and its performance is both fast and effortless [84]. Indeed, we know that small irreducible networks of delay-coupled neurons can behave as nonlinear filters, operating via competitive (nonlinear dynamical) modal resonances, as relays, switches, and propagators of possibly many different types of input-output patterns [64]. Such concepts are sympathetic to new computing paradigms. ANNs, with their simplistic weighted sums and imposed thresholds, bear no resemblance to the dynamic and nonlinear processing that we see in brain function (from high resolution fMRIs and in their performance). This makes the neural analogue for ANNs laughable, and wise proponents have abandoned it (but not the vocabulary).

On Partial Identity Verification

Identity, and its attribution and verification, is a very big deal within the digital society. There is often talk of obtaining absolute proof of identity, for example to access bank and other accounts. There is also the idea of the digital fingerprints: everything you do online leaves traces, and if these elements could be reassembled then could we infer your presence? There is a middle ground though, where we

continued

On Partial Identity Verification *Continued*

may not wish absolutely to identify an individual, but we may wish to estimate some partial aspect of identity. For example, on social media or dating sites it may be useful to automatically attribute a plausible age and gender to new friends. In fact, as platform operators including MNOs and websites move towards targeted marketing, they may need to demonstrate to regulators that they have taken care to asses such partial identities so as to supply appropriate messages and material to qualified, yet unknown, users. The methods in this chapter will assist with this, as will working with relatively small amounts of data in real-time.

Many companies would like to assess the behavioural mode that individuals are presently in. For example, if you are online or your mobile is on the network, we may wish to assess whether you are in work-mode, leisure-mode, family-mode, private-mode, travel-mode, resting-mode, party-mode, and so on. Depending on that attribution, you may be sent appropriate information or promotions. This is partial and softer than any absolute identification, and it is transient, unlike age and gender. Yet it is also more subtle than it seems.

In recent months I have been thinking about applications that might sample peoples' behaviour on a mobile network or online, and estimate whether they are doing things that are usual or unusual for them. The problem is that the usual and unusual both break down into a number of modes. In a security scenario we might have many people who are all accessing the same IT system, and each of them could have a distribution of their sampled activity defined over some space of all possible tasks. Then what is normal for one person may be unusual for another. So we might flag-up instances where individuals, acting within their own privileges, are nevertheless doing things, or doing sequences of things that are unusual for them. We might consider whether this is in response to some external events or otherwise motivated.

On Research Strategy

An interesting problem that besets almost all research programmes, whether publicly funded such as those of research councils and institutes, or privately funded such as those of corporates and investors, is that of achieving a balance between requirements-led activities and curiosity-led, radical, and possibly game-changing activities.

If you talk to your stakeholders (users, exploiters, or customers), they tend to suggest incremental improvements to the current paradigm in which they operate. They address their known unknowns, optimize their operations, and increase their chances of success. Yet the unknown unknowns are lurking out there. When something novel or radical happens, individuals, companies, or nations have to scramble to meet some new challenge that they have not prepared for. At such times, they cannot rely on their effectiveness within their current paradigm and they must seek out new concepts, ideas, and actionable methods. To be agile, it is a strategic necessity that research portfolios contain some curiosity-led research that they can call on when the game changes and some open-innovation partners who may have different skill sets, cross disciplinary experience, scientific know-how, and resources.

Sometimes the emergence of some new science is enough to change the game itself. The cavalry never asked for tanks, they merely wished for faster horses that would eat less. So some fraction of research programmes must invest in novel and radical ideas that might be game-changers or that could become essential. This science is both a high risk bet and an insurance policy. It does not need to be justified through current or foreseen economic and social activities.

Too often though, especially in times of tight funding, the research portfolio is pared down. This does not preclude future-looking themes or impactful research. It does mean that investments are justified in terms of known aspirations, market impacts, and so on. This tends to induce a group-think between the funders and their successful researchers. And when game-changing technologies or behaviours burst onto the scene, they all appear flat footed. A popular approach often adopted by governments is that of 'horizon scanning', which is a formalized process attempting to pick emergent challenges and technologies. The pace of change of both science and technology and its uptake means this is always out of date though. It also gives the illusion of thoroughness by filling up reports and slides. It is a sham.

These problems apply to analytics as well as everything else. I believe strongly that, like entrepreneurs, research portfolios should invest in the best people. We should do some things in maths and science because we have people with the vision, ability, and the passion to do them. We should not make everything managed and focussed. Perhaps twenty-five percent of a research portfolio should be aimed beyond the current paradigms and beyond any articulated requirements. Do public and private research programmes do this? Or do they play safe and sensible until the game changes?

I have served on two national research councils in the UK. I think the most important grants that we award are early-to-mid career fellowships. These are truly transformative for the individual recipients. They give them encouragement, esteem, resources, and time to think. When we come to thematic managed programmes, these often result in large grants that are poorly-managed and create a comfort zone where the investigators do not deliver on their promises on either the outreach or the sustainability of their centre or activity.

CHAPTER 6

Adaptive forecasting

6.1 Introduction to forecasting

S uppose that we wish to forecast the response to an event (such as the take-up of some offer or promotion, or a demand profile): we know exactly when the event starts, and we know roughly what the response should look like, but we do not have many actual data points. As we gain more and more early or recent observations we wish to forecast the future response, and we wish our forecasts to adapt and *tune in*.

We shall use the Bayesian updating framework set out in the Appendix. In particular, we will employ a form of (A.7), defined for continuous parameter vector $\mathbf{y}$, yielding a posterior distribution of $\mathbf{y}$, denoted by $P_{\text{posterior}}(\mathbf{y}|E, X)$, after observing a particular event E. It is given in terms of a prior distribution, denoted by $P_{\text{prior}}(\mathbf{y}|X)$, for $\mathbf{y}$, and a *model* term, denoted by $P_{\text{model}}(E|\mathbf{y}, X)$, which simply gives the probability of observing E, assuming any particular alternative values for $\mathbf{y}$. We have (from (A.7), in the Appendix)

$$P_{\text{posterior}}(\mathbf{y}|E, X) \propto P_{\text{model}}(E|\mathbf{y}, X).P_{\text{prior}}(\mathbf{y}|X).$$

Here, just as in the Appendix, X stands for 'everything else we know that might be relevant *a priori*'. Notice that we will not need to keep normalizing our distributions so that they integrate to one. Instead, we shall be taking logs and looking for local (and global) maxima, so any normalization will play no role.

Indeed within applications, such as those here, our work will involve making a sensible (subjective) choice for the $\mathbf{y}$-parameterized model term, and then choosing a value for $\mathbf{y} = \mathbf{y}^*$ to use in a future forecast. Almost always one

Mathematical Underpinnings of Analytics. First Edition. Peter Grindrod.
© Peter Grindrod 2015. Published in 2015 by Oxford University Press.

would choose a modal value for $\mathbf{y}^*$ at which the posterior has a local or global maximum. If the prior is uniform, this approach corresponds to maximum likelihood estimation, maximizing the probability of the evidence actually observed, with respect to the unknown parameters. Depending on the nature of the model, this task may require calculus or some numerical optimization procedure (either gradient or non-gradient methods) to find a modal value, $\mathbf{y}^*$.

6.2 Forecasting During Product Launches From Sales Data

Let $s^*(t|\lambda)$ be a time-dependent function for the level of response for $t \geq 0$, which is dependent upon a number of parameters held in a vector, λ, of non-negative parameters. For example, for a launch of a product or service we might set

$$s^*(t|\lambda) = \lambda_3 \frac{t^{\lambda_1}}{\lambda_2 + t^{\lambda_1}}, \quad \lambda = (\lambda_1, \lambda_2, \lambda_3).$$

Here, λ_3 represents the long-term response or sales rate while the short time response is governed by t^{λ_1}/λ_2. Alternatively for the consumer or market reaction (or its Twitter reaction) to the release of a some new movie, album, or book, we might propose various Poisson distributions, the simplest being

$$s^*(t|\lambda) = \lambda_1 e^{-t\lambda_2}.$$

So if all these unknown parameters were actually known we would have a plausible response profile, and we could exploit it to forecast the future response.

Given some observations we shall use s^* as a median forecast, and we shall estimate the λ parameters and simultaneously also estimate a confidence range about that forecast, accounting for the level of observed errors.

In practice, any observations, (s_i, t_i), of responses at various times will also have some errors as the formula will not be exact, so at any time t we will have to postulate that the actual response (which is assumed to be positive) is distributed about a median value that is given by $s^*(t)$. Perhaps the simplest assumption here is to assume it is log-normally distributed about $s^*(t)$.

So now we have a probability distribution of sales s at time t conditional on t, λ, and one extra positive parameter, σ, given by

$$P_{\text{model}}(s|t,\lambda,\sigma,X) = \frac{\exp\left(-\frac{\log(s/s^*(t|\lambda))^2}{2\sigma^2}\right)}{s\sigma\sqrt{2\pi}}.$$

For each fixed values of t and the parameters (λ,σ) we obtain a log normal distribution with median (and 95% confidence range) given by

$$s_{med}(t) = s^*(t|\lambda), \quad (s^*(t|\lambda)\exp(-1.96\sigma), s^*(t|\lambda)\exp(+1.96\sigma)).$$

Suppose that we guess some ranges for the parameters (this is called subjective expert judgement). We do so by not ruling out any values that could be feasible. This is relatively easy for the λ parameters, and σ should be less than $O(10^{-1})$, since if σ is much bigger we will be making errors in $\log s$ of $O(1)$, which would mean s^* is out and a factor of three and that would be rather useless. So if s^* is to be of any use at all, we will find $\sigma \sim O(10^{-1})$ at worst.

So let $P_{\text{prior}}(\lambda,\sigma|X)$ be a prior distribution for these: say, a uniform distribution, equal to unity (allowing an improper distribution) for all admissible values and zero otherwise.

Then suppose we have K observations $\{s_k,t_k\}$ (K might be quite small). Assuming these are independent (which they aren't really), we can write the posterior distribution, given by

$$P_{\text{posterior}}(\lambda,\sigma|\{s_k,t_k\},X) = \left(\prod_{k=1}^{K} P_{\text{model}}(s_k|t_k,\lambda,\sigma,X)\right) P_{\text{prior}}(\lambda,\sigma|X).$$

In practice it is better to take logs and to assume a uniform prior, and thus to maximize the posterior log likelihood (within the admissible prior ranges):

$$Q(\lambda,\sigma) = \sum_{k=1}^{K} \log(P_{\text{model}}(s_k|t_k,\lambda,\sigma,X)).$$

Note that we do not need to bother to normalize either the prior or the model distributions, since such constants will play no role here (justifying laziness). On the other hand if, as domain experts, we have a preference for some particular values for (λ,σ), then we can retain the prior term and just select a distribution (not necessarily normalized) with a maximum at our preferred point, and add the log probability term onto Q. This ensures that even with no data we can find a modal value for the log of the posterior distribution.

Then by brute force (for example, search, Monte Carlo, or random sampling), or better by gradient optimization we can find the maximizing (modal) values for (λ, σ), and we can employ those values in the above formula to estimate the future sales, $s_{med}(t)$, and the corresponding confidence range, at any future time, t. This is cheap, reliable, and robust. Code it up. Then argue about using alternative $s^*(t|\lambda)$s or priors.

Each time you add a new most recent observation, the maximizing value of the posterior will change and thus the subsequent forecast will respond. Eventually things will settle down and σ represents the irreducible volatility (the standard deviation of the log) observed about the median, $s^*(t|\lambda)$.

In Figure 6.1 we illustrate this with an example where we add successive data points and recalculate the modal values for the parameters at each step. In this case, we assume uniform prior over a set of admissible values and deploy $s^*(t|\lambda) = \lambda_3 t^{\lambda_1}/(\lambda_2 + t^{\lambda_1})$. The median forecast converges, as the mode converges towards $\lambda = (2.75725, 310.779, 1455.17)$ and $\sigma = 0.179419$, which is $O(10^{-1})$ as hoped.

Exercise 6.1

In Figure 6.2 we show a normalized plots of data from Google trends of weekly Google queries from in the UK, using the search 'cold+flu'. Typically there are two peaks: one after the return to school in September (when children get together in large numbers), with the peak and its duration dependent on the weather and other externalities; and one following Christmas (and the New Year return to work). Suggest a suitable model formulation that could forecast these profiles forewords each year from the first few weeks in June onwards, with the aim of predicting the sizes and timings of the peaks.

6.3 Suggested Project Work

Go to the website <http://www.oup.co.uk/companion/analytics> and access the file of new product launch data, NPL.txt. Notice that products not only vary in their total sales, but also by the number of outlets in which they are stocked. So the forecasts need to depend on time and the number of outlets. Produce plots similar to those in Figure 6.1. Give an overall estimate for the accuracy of such forecasts after three, five, and ten weeks of data, in predicting the future sales at Week fifteen.

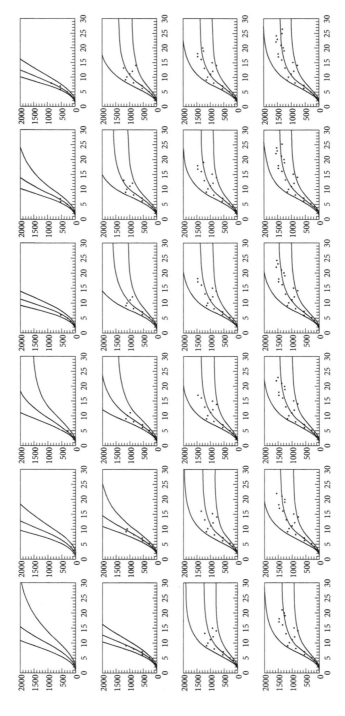

Fig 6.1 A product launch: median (rolling) forecasts and 95% confidence ranges evolve as new observations are made. Sales versus successive time intervals.

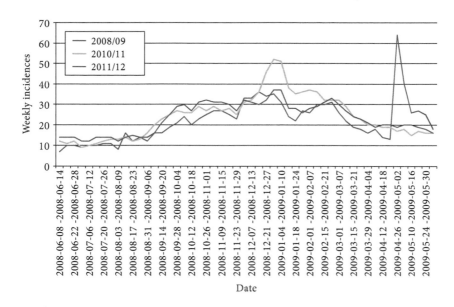

Fig 6.2 Normalized UK weekly incidences of the search 'cold+flu' from Google Trends (the anomalous spike in summer 2009 was the swine flu pandemic).

6.4 Exploiting Linearized Models

A very large class of models are intrinsically linear, or linearizable, and given in terms of covariates that are functions of the observable variables. Conceptually we assert and exploit the fact that errors in such models are distributed by a Gaussian. The *central limit theorem* says this must be the case if errors are made up from a sum of a large number of fluctuations, each drawn independently from any underlying distribution of finite variance. Moreover, as we shall see below, if in addition we assume a Gaussian prior, (which is conjugate, see Appendix, to the Gaussian model term) this results in a Gaussian posterior. In that case the subsequent optimization (finding the modal value for the posterior) requires the maximization of a quadratic form, which yields to calculus. Such models are the models of first resort: they are efficient, easily understood, and often intuitive. So let us write these models down explicitly in this section.

Moreover, below we shall discuss the case where we assert a prior distribution for both the unknown model parameters and the size of the model errors. By doing so we obtain a nonlinear equation to be optimized rather than the linear (normal) equations that are solved in standard linear regression (least squares), where it is usual to make no prior assertion whatsoever about the size of errors, and thus live with whatever is determined, regardless of plausibility. This extension allows us to start up with little or no data, and develop adaptive forecasts as more and more observations arrive.

Suppose that we observe some experimental data, $\mathbf{x}_i \in \mathbb{R}^n$ for $i = 1, \ldots, N$, from some real system, which we assume, for our own expert reasons, to satisfy a generalized linear equation:

$$y(\mathbf{x}) = \sum_{k=1}^{K} a_k(\mathbf{x}) \varepsilon_k.$$

Here the dependent variable, y, and the covariates, a_ks, are all given functions of the observable variables, $\mathbf{x}$, and the parameters, $\varepsilon = (\varepsilon_1, \varepsilon_2, \ldots \varepsilon_K)^T \in \mathbb{R}^K$, are to be determined. In practice there will be some errors in our deterministic model, and we should determine ε so as to minimize these. Typically y will involve some degrees of freedom in $\mathbf{x}$ beyond those required to evaluate the a_ks, in which case we might employ such a model to forecast future values of y, when we have yet to observe $\mathbf{x}$ in full, yet we can be explicit about those terms required to determine the a_ks.

Now suppose that we observe $\mathbf{x}_i \in \mathbb{R}^n$ for $i = 1, \ldots, N$ and we write

$$y(\mathbf{x}_i) - \sum_{k=1}^{K} a_k(\mathbf{x}_i) \varepsilon_k = e_i, \tag{6.1}$$

where the errors, e_i, are assumed to be independently normally distributed about zero with variance given by σ^2.

We let $\mathbf{y} = (y_1, y_2, \ldots, y_N)^T$ and let A be the $(N \times K)$ *design matrix*, with the ith row given by the row vector $\mathbf{a}_i = (a_1(\mathbf{x}_i), a_2(\mathbf{x}_i), \ldots, a_K(\mathbf{x}_i))$. Set $\mathbf{e} = (e_1, e_2, \ldots, e_N)^T$. Then, in vector notation, (6.1) becomes

$$\mathbf{y} - A\varepsilon = \mathbf{e}. \tag{6.2}$$

Our model is completed by making a hypothesis that the e_i are independent and normally distributed about zero with variance given by a further parameter $\sigma^2 > 0$.

We therefore have the conditional probability for the observed data, $(A, \mathbf{y})$, given the parameters (ε, σ):

$$P((A, \mathbf{y}) | (\varepsilon, \sigma), X) = \frac{1}{(2\pi\sigma^2)^{N/2}} \exp\left(\frac{-(\mathbf{y} - A\varepsilon)^T(\mathbf{y} - A\varepsilon)}{2\sigma^2}\right). \tag{6.3}$$

It $K \ll N$ then we expect a standard regression method will be useful, but the setting we have in mind here is that of making models as N increases. If we simply maximize $P((A, \mathbf{y})|(\varepsilon, \sigma), X)$, we obtain the usual normal equations and solution:

$$\varepsilon = (A^T A)^{-1} A^T \mathbf{y}, \quad \sigma^2 = (A\varepsilon - \mathbf{y})^T (A\varepsilon - \mathbf{y})/N.$$

This is clearly problematic when $N < K$ and $A^T A$ is singular, though pseudo-invertible with the SVD.

Now we consider our prior knowledge, denoted thus far by X. Before any data has been observed, and on the basis of our modelling experience, suppose that we expect that the parameters ε will be close to some given, preferred value ε_0. Then we may choose to employ a covariance matrix $\mathbf{C}$ to represent the acceptability of all other possible values, and hence our prior information is given by:

$$P_{\varepsilon \text{ prior}}(\varepsilon|X) = \frac{1}{(2\pi)^2 |\mathbf{C}|^{1/2}} \exp\left(-\frac{1}{2}(\varepsilon - \varepsilon_0)^T \mathbf{C}^{-1}(\varepsilon - \varepsilon_0)\right). \quad (6.4)$$

In O'Hagan and Forster (2004) [109] the use of this distribution is suggested, on the grounds that it is the conjugate of that in (6.3) with respect to ε. This simply means that when (6.3) and (6.4) are multiplied together, the result is a distribution (the factor in the posterior governed by ε of the same general form as (6.4)). Specifically we obtain:

$$P((A, y)|(\varepsilon, \sigma), X).P_{\varepsilon \text{ prior}}(\varepsilon|X) \propto \exp\left(-\frac{1}{2}(\varepsilon - \varepsilon^*)^T V(\varepsilon - \varepsilon^*)\right), \quad (6.5)$$

$$\text{where } V = \left(\frac{A^T A}{\sigma^2} + C^{-1}\right) \text{ and } \varepsilon^* = V^{-1}\left(\frac{A^T \mathbf{y}}{\sigma^2} + C^{-1}\varepsilon_0\right). \quad (6.6)$$

For a given σ, the mode and the expected value for ε, are thus coincident at ε^*.

Finally, having introduced σ, we must assert some acceptable prior distribution for its possible values. The variance σ^2 cannot realistically be as large as the variance in the (yet to be) observed values in $\mathbf{y}$. So, for example, we might assert that σ is distributed by a log normal distribution, with $\ln \sigma$ having an *a priori* expected value, μ_σ (say 1/3 of a typical value for $\ln(var(\mathbf{y}))$, and set $\mu_\sigma = \frac{\ln(var(\mathbf{y}))}{3}$). By asserting a standard deviation, η, for the normal distribution of $\ln \sigma$ equal to 3, we may allow this assumption about μ_σ to be in error by an order of magnitude or so either way. Hence we might choose:

$$P_{\sigma\ prior}(\sigma|\mathbf{X}) = \frac{1}{\sigma}\frac{1}{\sqrt{2\pi\eta^2}}\exp\left(-\frac{(\ln\sigma-\mu_\sigma)^2}{2\eta^2}\right), \tag{6.7}$$

where the constants μ_σ and η are known.

Now we may apply Bayes' rule with $\boldsymbol{\theta} = (\boldsymbol{\varepsilon}, \sigma)$. The joint posterior distribution for $\boldsymbol{\varepsilon}$ and σ is thus given by the multiple of the three distributions in (6.3), (6.4), and (6.7).

Rather than deal with this posterior distribution directly, we shall derive the conditions which determine the posterior modal values for $\boldsymbol{\varepsilon}$ and σ.

Forming the product, taking logarithm, and partially differentiating separately with respect to ε and σ, we obtain directly:

$$\varepsilon = (A^T A + \sigma^2 C^{-1})^{-1}(A^T \mathbf{y} + \sigma^2 C^{-1}\varepsilon_0), \tag{6.8}$$

$$0 = \mathbf{e}^T\mathbf{e} - \sigma^2 N + \sigma^3 \frac{d(\ln(P_{\sigma\ prior}(\sigma|X)))}{d\sigma}. \tag{6.9}$$

The first of these, (6.8) (exactly as in (6.6)), is a kind of generalization of the usual normal equations obtained in a standard regression analysis: as the data becomes more prevalent the first terms in each of the brackets will grow, while the terms in σ^2 will hopefully converge. However (6.8) remains valid for little or no data; regardless of whether $\mathbf{A}^T.\mathbf{A}$ is invertible, or usefully pseudo-invertible (via the SVD).

The error term $\mathbf{e}$ in (6.9) depends directly on ε: via $\mathbf{y} - \mathbf{A}.\boldsymbol{\varepsilon} = \mathbf{e}$. So by substituting for $\boldsymbol{\varepsilon}$ from (6.8) into (6.9), we obtain a single nonlinear equation for σ, which we may solve easily, by bisection for example.

If we adopt (6.7) as the prior for σ, then (6.9) becomes

$$0 = \mathbf{e}^T.\mathbf{e} - \sigma^2 N - \sigma^2 \left(1 + \frac{(\ln\sigma - \mu_\sigma)}{\eta^2}\right). \tag{6.10}$$

Notice (see O'Hagan and Forster (2004) [109]) from (6.9) that if our prior estimate for the modal value for σ happens to be correct, or if the prior for σ is a constant, and we use an improper prior reflecting our indifference (as in linear regression), then the last term in (6.9) vanishes and the modal σ^2 is simply the variance of the observed errors.

From (6.6) the modal value for ε is also the expected value, and we can see that the covariance matrix in the distribution for ε is simply the inverse of $\mathbf{V}$, so further information about the likely values for ε is at hand.

Now consider the scenario where we make a model and adapt it each time the number of observations N increases. With no data we estimate $\varepsilon = \varepsilon_0$ and $\sigma = e^{\mu_\sigma - \eta^2}$. Notice this last is the modal value of the prior distribution we assumed for σ, and not the expected value.

Each time we observe a new pair $(\mathbf{a}_{N+1}, y_{N+1})$, from a new $\mathbf{x}_{N+1}$, we may extend both A and $\mathbf{y}$, increment N to $N+1$, and recalculate any required forecasts, y, corresponding to various proposed values for $\mathbf{a}$.

In practical applications it may be that one component of $\mathbf{a}$ is a constant (equal to one), unless we are sure that $\mathbf{a} = 0$ implies $\mathbf{y} = 0$, and also that one or more of the components may depend upon time, t (that is observed without error and included as part of $\mathbf{x}$).

One assumption that we have made implicitly above is that aged observations are as important as more recent observations: all errors are penalized in the same way. In the next section we shall encounter a situation where we build no-in a higher tolerance to aged errors, compared to more recent errors, in optimizing a forecast. This is equivalent to weighting the observations using one of the observables (the age of the observation). This idea could of course be incorporated in the current generalized linear model setting here, and is left as a project for the interested reader.

Exercise 6.2

Show that the model term

$$P((A, \mathbf{y})|(\varepsilon, \sigma), X) = \frac{1}{(2\pi\sigma^2)^{N/2}} \exp\left(\frac{-(\mathbf{y} - A\varepsilon)^T(\mathbf{y} - A\varepsilon)}{2\sigma^2}\right)$$

is maximized at

$$\varepsilon = (A^T A)^{-1} A^T \mathbf{y}, \quad \sigma^2 = (A\varepsilon - \mathbf{y})^T(A\varepsilon - \mathbf{y})/N,$$

when $(A^T A)^{-1}$ exists.

Exercise 6.3

Show that (6.5) holds where V and ε are given by (6.6).

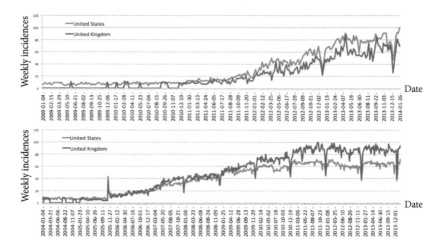

Fig 6.3 Normalized US and UK weekly incidences of the search 'Big Data' (above) and 'Analytics' (below) from Google Trends (the spike in 'Analytics' is the launch of Google Analytics in November 2005).

Exercise 6.4

In Figure 6.3 we show a normalized plots of data from Google Trends of weekly Google queries from the US and UK, for the terms 'big data' and 'analytics'. Suggest a suitable linear model formulation that could forecast these profiles forewords from the first few weeks onwards.

6.5 Forecasting Energy Demand From Smart Meter Data

In this section we move to a much more complicated, real life, example, but the principles remain the same. Besides asserting a more complicated model distribution for the actuals given the forecast (and parameters), this particular form is not entirely differentiable since it contains a minimization over a set of permutations (and with consequent penalties). So it requires a combination of calculus and a non-gradient method to find the mode, and hence generate the forecasts.

We consider smart meter data for domestic energy consumers. This is typically measured by smart meters at half hourly intervals. The problem we have in mind is that of forecasting spikes during each coming day. Rather than making a smooth median forecast and treating the spikes as errors, we wish to place our focus on the spikes. Moreover, since we may wish to control other hardware assets such as batteries on the local low voltage (240v in the UK) network, we want to have an early warning of such spikes in demand. We follow [22] throughout this section.

In Haben et al. (2013) [68] the authors introduced a novel error measure that rewards the forecasting of extremes (spikes) in behavioural profiles, by allowing for a relatively limited amount of time shifting, permuting the forecast to match the actual, possibly under some penalty. This is designed to reward forecasters that predict extreme sharp spikes, yet may get the precise timing of such events slightly wrong. For example, if a method estimates a sharp spike but the actual contains such a spike but at a slightly different time, then a regular norm penalizes that forecast twice, once for the forecast spike and once for the missed actual spike. Thus, when such simple error norms are used, it may have been better to forecast no spike at all and thus be only penalized once for the miss. The measure introduced by Haben et al. (2013) [68] and given below removes that disincentive.

Here, we shall consider a variation of that type of measure that can bias the consequent forecaster to predict early (rather than late) spikes. In some applications, where a warning or preparation time for such extremes is desired, such a conservative warning forecast is advantageous. In other applications, such as the smart control of local energy storage and its subsequent release, the forecaster-controller should not rely on low-demand periods immediately prior to a predicted high-demand spike to increase the storage demand (by charging up), since an early actual spike, even if unlikely, will result in an overall worse situation, possibly violating the available supply headroom.

This method may be useful in a wide variety of applications where the profile data to be forecast is spiky, and where early forecasting of spikes is desirable. In the example of household energy demand an efficient (low cost) and accurate (yet sometimes conservative) forecast methodology, allowing for rapid calibration (with as little history as possible) is highly desirable on both sides of the metre.

We introduce a Bayesian framework for deriving forecast based on minimizing a generalization of the error measure introduced in Haben et al. (2013) [68].

We wish to forecast a demand profile for a single day (or some other time period), where each forecast and each observation comprises of a vector of $m > 1$ ordered point measurements (for example, one point for each of the $m = 48$ half hours per day). Then we shall measure the error by making a direct comparison between the forecast vector and the actual observed profile for that day.

Let $\mathbf{f} = (f_1, \ldots, f_m)^T$ and $\mathbf{a} = (a_1, \ldots, a_m)^T$ denote a forecast and an actual observed profile, respectively. For standard domestic smart meters this is typically $m = 48$ (every half hour). However, for higher resolution monitoring in other applications we may set m much larger. Then

$$E(\mathbf{f}, \mathbf{a}) = \min_{S} \left(\sum_{i=1}^{m} |f_{S^{-1}(i)} - a_i|^p + a_i^p g(S^{-1}(i) - i) \right)^{1/p}, \qquad (6.11)$$

denotes the error measure. Here S ranges over all suitable permutations of the index set $\{1, 2, \ldots, m\}$. If $\mathbf{f}$ has a peak at $j = S^{-1}(i)$, which is matched by S to a corresponding peak, say a_i, at $i = S(j)$, then if $j < i$ the peak is forecast early, which can be much better than when $j > i$, and it is forecast late. So here the shifting-penalty function, $g(j)$, must be such that

(i) $g(0) = 0$,
(ii) g increasing for $j > 0$,
(iii) g decreasing for $j < 0$,
(iv) $0 < g(-j) < g(j)$ for any $j > 0$.

The parameter $p \geq 1$ biases the measure to penalize errors at peaks: typically we shall take $p = 4$. The calculation of $E(\mathbf{f}, \mathbf{a})$ thus requires a fast search over permutations, S, that can be achieved efficiently using the (very efficient) Hungarian algorithm, or some similar method [68], [23]. The Hungarian algorithm produces pairwise matches between a set of m forecasts and m actuals so as to minimize the sum of the costs of the pairwise matches. It has an interpretation in terms of bipartite graphs.

Now suppose we are given a set of K actuals (historical daily observations) given by vectors $\{\mathbf{a}_k | k = 1, 2, \ldots, K\}$, together with the corresponding exact *ages* for each observation, without loss of generality in increasing order, say $\{t_k | k = 1, 2, \ldots, K\}$. We wish to use these to condition a single-day forecast, $\mathbf{f}$, for use at the next relevant day, where we may place relatively more weight on the influence of relatively recent actuals (with relatively lower ages).

In our present notation, Bayes' theorem (see the Appendix) says

$$P(\mathbf{f}|\{\mathbf{a}_k, t_k\}_{k=1}^K, X) \propto P(\{\mathbf{a}_k, t_k\}_{k=1}^K|\mathbf{f}, X).P_{prior}(\mathbf{f}|X). \tag{6.12}$$

As usual the middle 'modal' term describes the distribution of errors (from the actuals), given the forecast. This term may contain some other parameters $(\sigma, \lambda, \ldots)$, see below, which must be determined along with the forecast $\mathbf{f}$. So we can extend equation (6.12) to write

$$P((\mathbf{f}, \sigma, \lambda)|\{\mathbf{a}_k, t_k\}_{k=1}^K, X) \propto P(\{\mathbf{a}_k, t_k\}_{k=1}^K|(\mathbf{f}, \sigma, \lambda), X).P_{prior}((\mathbf{f}, \sigma, \lambda)|X).$$

The age t of an observation is known exactly; it is the observations that may differ from the forecast $\mathbf{f}$. We shall consider the following negative exponential model for a single daily observation $\mathbf{a}$ with age t:

$$P((\mathbf{a}, t)|(\mathbf{f}, \sigma, \lambda), X) = e^{-E(\mathbf{f}, \mathbf{a})\lambda^t/\sigma}\frac{\lambda^t}{\sigma}. \tag{6.13}$$

Here, $\lambda \in (0, 1]$ is an age-discounting factor, and $\sigma > 0$. The larger the observation's age, t, the larger the expected value (and variance) of the distribution for the error $E(\mathbf{f}, \mathbf{a})$. For the negative exponential distribution we have a mean error given by $\sigma/\lambda^t \to \infty$ and an error variance given by $\sigma^2/\lambda^{2t} \to \infty$ as $t \to \infty$.

Now suppose that the observations are independent, then we can write

$$P(\{\mathbf{a}_k, t_k\}_{k=1}^K|(\mathbf{f}, \sigma, \lambda), X) = \exp\left(-\sum_{k=1}^K \lambda^{t_k} E(\mathbf{f}, \mathbf{a}_k)/\sigma + \ln\lambda \sum_{k=1}^K t_k - K\ln\sigma\right). \tag{6.14}$$

Assuming for the moment that the prior is uniform and plays no role, we will maximize the posterior distribution (6.14), which is equivalent to finding maximum likelihood estimates (MLEs) for the unknowns $(\sigma, \lambda, \mathbf{f})$ that jointly maximize the model for the aged observations.

We must therefore minimize the negative of the log-likelihood, denoted by,

$$H(\sigma, \lambda, \mathbf{f}) = \sum_{k=1}^K \lambda^{t_k} E(\mathbf{f}, \mathbf{a}_k)/\sigma - \ln\lambda \sum_{k=1}^K t_k + K\ln\sigma. \tag{6.15}$$

Now differentiate H with respect to σ to obtain

$$0 = H_\sigma = -\sum_{k=1}^K \lambda^{t_k} E(\mathbf{f}, \mathbf{a}_k)/\sigma^2 + K/\sigma,$$

so that

$$\sigma = \frac{1}{K} \sum_{k=1}^{K} \lambda^{t_k} E(\mathbf{f}, \mathbf{a}_k).$$ (6.16)

Now differentiate H with respect to λ to obtain

$$0 = H_\lambda = \sum_{k=1}^{K} t_k \lambda^{t_k-1} E(\mathbf{f}, \mathbf{a}_k)/\sigma - \sum_{k=1}^{K} t_k/\lambda.$$

Thus we must have

$$\sigma = \frac{\sum_{k=1}^{K} t_k \lambda^{t_k} E(\mathbf{f}, \mathbf{a}_k)}{\sum_{k=1}^{K} t_k}.$$ (6.17)

From (6.16) and (6.17) we see that λ must satisfy

$$Q(\lambda) \equiv \sum_{k=1}^{K} \lambda^{t_k} E(\mathbf{f}, \mathbf{a}_k)(t_k - \bar{t}) = 0,$$

where $\bar{t} = (t_1 + \cdots t_K)/K$ is the mean observation time. Clearly $Q(0) = 0$ and $Q(\lambda) \sim \lambda^{t_1} E(\mathbf{f}, \mathbf{a}_1)(t_1 - \bar{t}) < 0$ for small λ. Similarly for large λ we have $Q > 0$. Thus there is at least one root for $\lambda > 0$.

Hence if $\mathbf{f}$ is given, we may solve (6.16) and (6.17) together for $\sigma = \sigma(\mathbf{f})$ and $\lambda = \lambda(\mathbf{f})$. If the solution is such that $\lambda > 1$, then we must set $\lambda(\mathbf{f}) = 1$ (no discounting) and $\sigma(\mathbf{f})$ is given by (6.16), so in that case $\sigma(\mathbf{f}) = \frac{1}{K} \sum_{k=1}^{K} E(\mathbf{f}, \mathbf{a}_k)$. Finally, in all cases, we must search for $\mathbf{f}$ so as to minimize H.

Substituting from (6.16) into (6.15) we obtain a simplified objective

$$H(\sigma, \lambda, \mathbf{f}) = -\ln \lambda(\mathbf{f}) \sum_{k=1}^{K} t_k + K(1 + \ln \sigma(\mathbf{f})).$$ (6.18)

In deploying any discrete search algorithm, such as a genetic algorithm (GA) (see Section 7.3), or Nelder-Mead, for example, or other non-derivative optimization methods we may trial any suitable $\mathbf{f}$, calculate the errors

$$E(\mathbf{f}, \mathbf{a}_k) \quad k = 1, \ldots, K,$$

solve (6.16) and (6.17) together for $\sigma(\mathbf{f})$ and $\lambda(\mathbf{f})$, and then finally use these values in the objective function (6.18). If we obtain a maximum where $\lambda \to 0$, we are in effect using only the most recent (the least aged) observation as the forecast. If we obtain a maximum where $\lambda = 1$ then we are using all observations equally regardless of their age. For intermediate values of λ we can get some idea of how much history is really relevant to a forecast. For example, we could ignore observations for which $t_k > t_1 + 2/(-\log_{10} \lambda)$, where the observations have weights less than one per cent of the most recent observation.

The parameter σ measures the overall size of errors and is small for consistent and forecastable households, and large for inconsistent and volatile households.

It seems simplest to use uniform priors for $\lambda \in (0, 1]$ and $\sigma > 0$ but if K is small we may wish to have a prior for $\mathbf{f}$, so as the number of observations increases the prior estimate exerts less influence on the forecast. The easiest way is to assume

$$P_{prior}(\mathbf{f}|X) = \frac{e^{-E(\mathbf{f},\mathbf{f}^*)/\sigma_0}}{\sigma_0},$$

for some given $\sigma_0 > 0$ and prior forecast $\mathbf{f}^*$.

Then H (from the posterior) in (6.18) must be augmented with an extra term and becomes

$$H(\sigma, \lambda, \mathbf{f}) = -\ln \lambda(\mathbf{f}) \sum_{k=1}^{K} t_k + K(1 + \ln \sigma(\mathbf{f})) + E(\mathbf{f}, \mathbf{f}^*)/\sigma_0,$$

while equations (6.16) and (6.17) must still hold as before. Clearly for little or no data we will have $\mathbf{f} = \mathbf{f}^*$, the prior modal estimate. The last term may dominate depending on both K and σ.

Now we apply the forecast methodology described earlier to some real smart meter data from the Irish smart meter trial [82]. To define our error measure we take $p = 4$ in the error equation (6.11) to weight the model to favour larger peaks, and we define $g : \mathbb{Z} \to \mathbb{R}$ by

$$g(j) = \begin{cases} 0.05j, & \text{if } j \leq 0, \\ 0.1j, & \text{if } j > 0. \end{cases} \tag{6.19}$$

Thus we penalize late forecast peaks twice as much as early peaks.

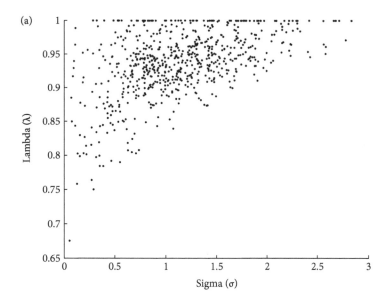

Fig 6.4 Relationship between σ and λ from Charlton et al. (2013) [22].

We forecast each day of a single week for one hundred domestic customers using the same day from the previous ten weeks as aged actuals, giving us 700 forecasts in total. We employ a uniform prior and find the maximum likelihood estimates $(\sigma, \lambda, \mathbf{f})$ using a genetic algorithm (see Section 7.3).

The model parameters σ and λ play an important role in determining the accuracy of the data and quantity of historical (aged) data used to create the forecast. Figure 6.4 shows the relationship between λ and σ for 700 household forecasts. In general, more variable customers (larger σ) tend to use more days of the history (have λ values closer to 1) in formulating the forecast.

The forecast outperforms two naïve forecasting methodologies: (i) a simple persistence forecast using the previous week as the forecast and (ii) a forecast employing the standard mean profile from the previous ten weeks. It scored lower adjusted errors, $E(\mathbf{f}, \mathbf{a})$, for 483 forecasts compared to the persistence forecast, and for 527 forecasts compared to the mean forecast. It often scored poorly for extremely volatile customers who had an exceptionally large or small usage on the forecasted day (as would any model).

The exemplars in Figures 6.5-6.6 shows the daily profiles of the ten historic weeks and then the final forecasted week, together with the actual at the bottom. The plots show some of the success of the method not only in forecasting the correct magnitude of the peaks, but also the correct approximate timing.

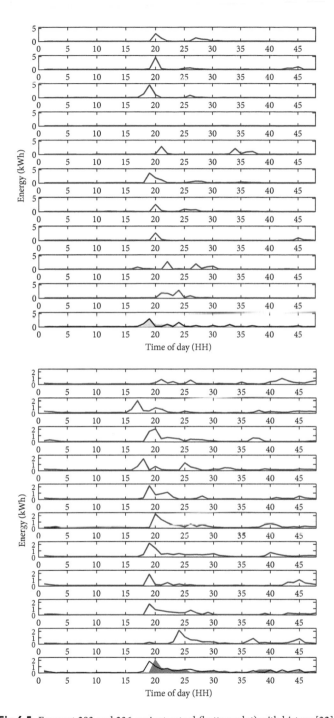

Fig 6.5 Forecast 202 and 236 against actual (bottom plot) with history [22].

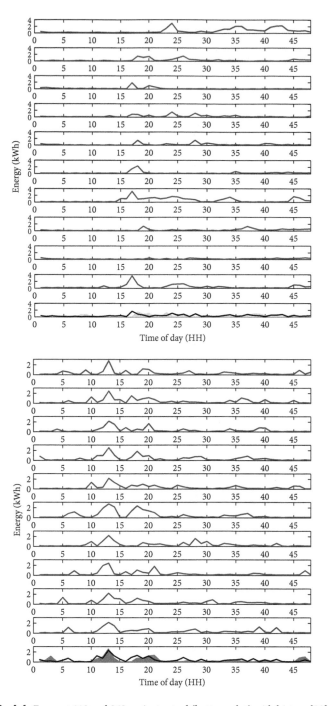

Fig 6.6 Forecast 112 and 268 against actual (bottom plot) with history [22].

In particular, the examples show the tendency of the MLE to favour forecasting early peaks with the historical data. For example, the historical data on the left of Figure 6.5 show a peak around the 20th half hour (10am) but due to the appearance of a peak prior to 10am in two of the historical profiles, the forecast has successfully anticipated an early peak for the forecasted week.

6.6 Suggested Project Work

Fetch the data from the Irish National Smart Meter Trial [82] and carry out some forecasting!

A Personal View

On Elusive and Naïve Forecasts

Some things that one feels intuitively ought to be forecastable turn out to be difficult to crack. Other things that seem like a long-shot yield readily. The spiky household energy demand data discussed in Solutions to exercise section are an example of the latter. Of course we have to specify what we want in a forecast, and in that case we wished to bias our forecast so as to yield early spikes. Although this introduced two layers of complication (the Hungarian algorithm to permute prototype forecasts to match actuals and a p-norm to penalize missed spikes), the principle was actually very clear. Of course the complication resulted in the need for a combination of discrete search and calculus.

An example of the former type is the impact for pricing and promotions. Demand theory suggests that individuals making rational decisions should generally show lower demand, d, for a product as the price, p, increases: $d \propto p^{-\gamma}$ for some exponent, γ, representing the elasticity. So, given the sales of a product (or exchangeable group of products) we might fit γ. Then revenue is $pd(p)$, while costs are related to the supply of d so the profits might be optimized. The problems with this are that some goods are apparently price independent in the normal range, or demand may even increase with price. People do not behave rationally: events such as pricing and promotions intervene, or their usage of a product is non-negotiable at the usual level. For example, there is a fixed number of domestic cats in the UK and they must each have a portion of cat food every day, so there is an irreducible need for a certain level of cat food. Pricing and promotions merely cannibalize future sales (stock-piling), so the true purpose of such activity is not to sell more cat food in the long-term. If you plot demand (say weekly) or

continued

On Elusive and Naïve Forecasts *Continued*

share of total category demand for products against price (or price relative to category), you see a scatter plot with a very poor demand-price curve. Fitted elasticities are often fictional. Even so, 'demand optimization' is intuitively accepted and expected by brand and category managers. Very few products have reasonable and reliable quantitative estimates for their elasticities, (just ten per cent in most categories).

Examples of negative elasticities include Giffen goods which see increased consumption as individuals' income rises, even as prices rise. They also occur through scarcity giving rise to a scarcity heuristic enabling us to make non-rational fast decisions, for example 'four other people are currently viewing this hotel', or else where unit prices are so low that consumers make a decision in the quality domain rather than the price domain. In the UK this happened ten years or so ago when 'sausage price wars' resulted in huge discounts on branded sausages (and also own-label sausages, generally at a lower price), which resulted in the consumer equating price with quality (reasonably asking 'What is in these sausages?'), and their willingness to pay showed demand for relatively higher-priced products.

The problem arises more clearly when one analyses promotions. Distinct types of promotions behave very differently. Buy-One-Get-One-Free is the same as fifty per cent off (half-price) in terms of the net unit price, p, to the customer, yet they behave very differently. I once carried out an analysis where we looked at the promotional volume uplift of many products in a supermarket chain (promotional sales rate divided by baseline sales rate) as a function of promotional attributes. These last included the price model (BOGOF, Buy-One-Get-Second-Half-Price, percentage reduction), store display options, and marketing support options, all as represented by categorical variables, as well as the entry price (minimum spend required to trigger the promotion), and the net unit price. This proved very useful indeed. In particular, it demonstrated the additional uplift achievable by employing secondary displays at the ends of aisles. It was clear that there was a 'power of free' operating. This has been written about in Ariely (2008) [6], and of course Kahneman (2011) [84], and is not new to behavioural psychologists, so we need to incorporate this into our modelling. The naïve demand-curve of microeconomics, and the consequent demand optimization paradigm, is simply not up to this task, and reveals little information. Of course once we understand this, we can get to work with suitable models. We cannot be successful in behavioural analytics without being open to such wisdom. Big data cannot make the wrong concept into a right one: there is no brute solution. Often I meet heads of customer insight within companies (or well-dressed folk with similar titles and remits) who seek to impress with the weight of their data. This is a pity, but most likely it reflects the challenge that they face in processing the data deluge from day-to-day, rather than their effort invested in finessing their insights, and thus their employers' competitiveness.

Similar issues pertain in space optimization problems. Suppose we have some extra space or bandwidth, for example page-space in a marketing flyer, or within an online landing page, or we have extra shelf units (called mods) within an aisle. What type of product or which particular brands or stock-keeping units (SKUs) should we place there?

The e-commerce side to this problem is growing. There is no reason why online stores should appear the same to all customers. Indeed, recommender systems (see Amazon's 'people who bought this book also bought...') recognize this. But thinking bigger, the website does not need to have the same look or feel to each browser. The online equivalents of customer mission and users' mode (see earlier chapters) might be inferred with a suitable segmentation, and some consequent functionality can be developed that is unique and differentiating.

There is a reason why some forecasts are elusive: either you are employing the wrong concepts, and thus the wrong models, or else the behaviour is genuinely random. The latter case is possible, as there are of course some customers who may behave randomly for most or all of the time, so we should pause and consider this hypothesis. However, the failure of a naïve forecast usually indicates that we should look elsewhere for some alternative modelling concepts, and by doing so we may discover some more effective and possibly very profitable insights.

CHAPTER 7

Customer journeys and Markov chains

7.1 Customer Behaviour and Value

Any modern business recognizes that its key asset is the customer base, and its key activity is interacting with the customer. Businesses can evolve their product and service offerings (sometimes organically, sometimes radically), but the value of the business to its investors and shareholders is based on what it will do and how it might perform in the future, and where its future revenues will be earned. For many, the customer base is large and requires management and planning. Do they acquire their most valuable customers or do they need to develop them?

Management science is alive to this and recent years have seen a massive development of strategic and business functions which includes the rise of customer relationship management (CRM) [65]. Businesses need to model, estimate, and manage their customer behaviour and response to future initiatives to grow *customer lifetime value* (CLV). There is a fairly recent account of this topic, and potential solutions, given in Gupta et al (2006) [66].

Customers' behaviour evolves over time as they respond to the business offering, build up transaction history and experience, and adjust to their own lifestyle demands. The changes in customers' behaviour drive changes in their future value to the business. In order to understand and define expectations for the future return from customers it is necessary to understand how future changes in behaviour can be inferred from their current behaviour. Behaviour

Mathematical Underpinnings of Analytics. First Edition. Peter Grindrod.
© Peter Grindrod 2015. Published in 2015 by Oxford University Press.

is defined by the way the customer transacts with the business and the type of transactions made over some given time period, say a month or so.

Calculations of CLV (defined as the future sum of value or profits over the customer's future life) vary according to the sector that a business is in Gupta et al (2004) [67]. They may employ some discounting of future revenues, or truncation at some suitable time horizon. Subscription services are particularly simple, whereas those situations where the time dependent customer value is highly behaviour-driven and highly variable are more difficult to model and manage.

This chapter focuses on the retail sector, for which the approach here was devised and applied, but we also present a case study related to mobile telephone company. The application of Markov chain models has also been advanced for similar reasons for businesses where customer states may be based on the recency of transactions [110], and also within the insurance sector [119].

7.2 Markov Chains

We begin with a definition for a discrete-time Markov chain, named after Andrey Markov [43]. Suppose that we have a population of entities, each of which is in exactly one of a finite number, n, of *states*, at successive times steps, $\{t_k\}$. If we denote the states by $i = 1, \ldots, n$, then the discrete time behaviour of each entity results in a sequence, or chain, $\{c_1, c_2, c_3, \ldots\}$, over time steps $\{t_1, t_2, t_3, \ldots\}$. In a Markov chain the sequences are defined stochastically, with the next state being conditionally dependent on the present state, but not on any further previous history. Thus, successive states are linked together in a chain-like fashion. We have probabilities of the form

$$P(c_{k+1} = j | c_k = i, X).$$

These are called (conditional) transition probabilities and are stored in an $n \times n$ matrix, called the transition matrix, A, say

$$A_{ij} = P(c_{k+1} = j | c_k = i, X) \quad i, j = 1, \ldots, n.$$

In this case the rows of A must sum to one. We shall write this more simply as a conditional probability distribution for the next state j, given the current state i:

$$A_{ij} = P(j | i, X) \quad i, j = 1, \ldots, n.$$

Then let $\mathbf{x}_k$ denote the distribution of a population of entities over the set of states at time t_k. Assuming that all individual entities' transitions are independent then the expected population at the next time step is given by

$$\langle \mathbf{x}_{k+1}|\mathbf{x}_k \rangle = A^T \mathbf{x}.$$

This is the Markov model. Together with the independence assumption, the matrix A contains all the information that we require to generate instances (sequences of successive states) of the Markov chain.

Typically in analytics we will be travelling in the inverse direction—given observed sequences, construct a suitable model.

If the assignments of the actual discrete time observations to the states is not known to us, and is assumed but hidden, then we must determine both a suitable definition for a set of states (in terms of observables) and the corresponding transition matrix.

Exercise 7.1

Given that $P(i_{k+1}|i_k, X) = (A)_{i_k i_{k+1}}$, what is $P(i_{k+2}|i_k, X)$?
Given that $\langle \mathbf{x}_{k+1}|\mathbf{x}_k \rangle = A^T \mathbf{x}_k$, what is $\langle \mathbf{x}_{k+2}|\mathbf{x}_k \rangle$?

Exercise 7.2

Suppose that A is irreducible and that we iterate $\mathbf{x}_{k+1} = A^T \mathbf{x}_k$. What is $\lim_{k \to \infty} \mathbf{x}_k$?

In order to calculate estimates for CLV based on behavioural change we shall proceed stepwise over discrete time steps. Within each time step we shall allocate each and every observed customer to exactly one of a discrete number of behavioural states. At the same time, we shall derive a Markov model for the state-to-state transitions: a conditional probability distribution for stepwise behaviour in the next time step, dependent on that at the current time step. There are a number of variations we might consider.

The simplest case is where we are given the states *a-priori* (defined for historical or business reasons). Then we shall need to calibrate a suitable Markov model, giving a good representation of the observed populations of all state-to-state transitions. For simplicity we might begin by assuming *stationarity* of the process. Stationarity means that the generative (stochastic) process is itself time invariant, and thus the parameters describing all probabilistic transitions are themselves constants (if known), so then we may consider any and all state-to-state transitions together, for all individuals and all successive time steps.

Alternatively, we might have a number of choices regarding possible state definitions, or search within a huge class of possible choices, each of which defines a partition of the behaviour observable within a single time step. A good partition will result in a model where the futures of individuals within distinct states are also relatively distinct, since if two or more states have the same conditional distributions over possible next states for the next time step, then they are not separated from a predictive point of view: we may as well just combine them. Similarly, we might wish the Markov model to be dominated by a few key, relatively high-probability state-to-state transitions. So we might require the transition matrix to be sparse.

A third alternative occurs where we might seek a hidden Markov model (HMM) using a maximum likelihood method to define the state-partition and the consequent transition probabilities, so as to make the observed sequences (from all historical behaviour) as likely as possible. This type of model often employs a variant of the EM algorithm (see Chapter 2) and is usually applied where the observed behaviour is itself categorical and thus of relatively low dimension. Then the use of multinomials for the conditional distribution for such observables keeps the calculation manageable. So in such circumstances this is a viable approach. We shall not discuss the HMM approach further here because typically within our applications we will observe a relatively large dimensional vector summarizing the behaviour of each customer within each time step. It is typically a mix of real and categorical variables, and not a single categorical variable, and business-user requirements will impose some constraints on the final model. Having the theoretically optimized Markov model may not be useful for business purposes, for example, if it contains state-populations of wildly different sizes, or if the transition matrix is ill-conditioned.

Here, we shall focus on the second alternative.

7.3 Genetic Algorithms

Since we may need to search for good models, we begin by discussing a popular, flexible, and inexpensive method of non-derivative optimization.

In artificial intelligence, a genetic algorithm (GA) is a heuristic search method that mimics the process of natural selection in optimizing an objective function. It is a non-derivative method, so the objective simply needs to be well defined, rather than differentiable. Derivative methods, based on search directions [111], require the existence of gradients and Hessians of the objective (with respect to the unknowns). A GA is an evolving search, and GAs are applied in many, many fields.

Although genetic algorithms can be applied blindly, simply coding up the problem, this is often sloppy and expensive. It is productive to apply the method sympathetically, in the light of prior knowledge and information about the likely form of the solution, so as not to make the computation orders of magnitude longer than necessary.

A genetic algorithm produces a *pool* (a set) of potential solutions at each of a number of successive generations, $k = 1, 2, \ldots$. Suppose we have an objective function $F(\mathbf{y})$ that is to be optimized by a choice of $\mathbf{y}$ in a space Y of possible solutions. Note components of $\mathbf{y}$ may be integers, categorical variables, or real numbers. For each k we will have a population, or pool, of $m \gg 1$ possible solutions:

$$\mathcal{P}_k = \{\mathbf{y}_i^{[k]} | i = 1, \ldots, m\}.$$

Each $\mathbf{y}_i^{[k]}$ (sometimes called a genotype) in the kth pool corresponds to a *fitness* measurement, given by the objective

$$F_i^{[k]} = F(\mathbf{y}_i^{[k]}).$$

We shall define a process of evolution so that the $\mathcal{P}_k$s evolves towards an ensemble that yields close to optimal values of fitness (that is, near-maximizing the objective).

We require that the potential solutions $\mathbf{y} \in Y$ may be combined and manipulated so as to produce other candidate solutions, also in Y. Specifically, we shall assume that there is a probability distribution function C over Y-space that is conditional on a pair of *parents*, in $Y \times Y$, denoted by

$$C(\mathbf{y} | \mathbf{y}_a, \mathbf{y}_b, X).$$

This is called the *cross-over and mutation* function. Given the pair $(\mathbf{y}_a, \mathbf{y}_b)$ we can generate offspring by sampling from the distribution C. Sensible constraints are that C be symmetric,

$$C(\mathbf{y}|\mathbf{y}_a, \mathbf{y}_b, X) = C(\mathbf{y}|\mathbf{y}_b, \mathbf{y}_a, X),$$

and provides some dispersion around spliced, random, combinations of the parents. In particular,

$$C(\mathbf{y}|\mathbf{y}_a, \mathbf{y}_a)$$

should have a modal value at $\mathbf{y}_a$ and a relatively small dispersion about that sole parent.

Traditionally, this is carried out computationally by employing a binary vector as a parameterization of Y-space, so all solutions are just binary strings, and thus $Y = \{0, 1\}^s$ for some integer $s \gg 1$. Then independent draws from C (which is only defined implicitly in practice) are made via a two-step processes: (i) *cross-over*, where the two parent strings are randomly spliced together to produce a candidate offspring and (ii) *mutation*, where each and every resultant binary element is flipped with some small independent probability.

For crossover we simply generate a number of splicing positions, $\{s_1, s_2, \ldots, s_k\}$, such that $1 = s_1 < s_2 < \ldots s_k = s+1$ and then for each substring from s_r up to $s_{r+1} - 1$ we insert the corresponding substring of binaries from either $\mathbf{y}_a$ or $\mathbf{y}_b$, selecting one of the parents at random (or alternating them after a random start). The result is a randomized hybrid string, with each complete substring drawn from one or other of the parents. It is clear that if a single parent is crossed with itself then the result is a copy of the single parent.

Mutation is more trivial: starting from the candidate offspring from crossover, simply flip each element independently ($y_j \rightarrow 1 - y_j$) with a small probability p. Note that if s is large then the probability of no mutation occurring is $(1 - p)^s$. So p needs to be thought about in terms of the size, s, of the candidate solution. Typically one could wish for $(1 - p)^s = 0.95$ or so.

In practice, one may be a little more pragmatic about crossover in particular, which is clearly related to the parameterization of Y-space. Some care and experimentation needs to be taken.

Now we can run the GA. In addition to C (equivalent to p and the other splicing details) we select a pool size, $m \gg 1$, and a real parameter $\mu > 0$.

At $k = 1$ we generate a pool $\mathcal{P}_1$ containing m independent guesses $\{\mathbf{y}_i^{[1]} | i = 1, \ldots, m\}$ drawn from Y.

Now we define the evolving step. Given $\mathcal{P}_k = \{\mathbf{y}_i^{[k]} | i = 1, \ldots, m\}$ for $k \geq 1$ we proceed as follows.

(a) For each $i = 1, \ldots, m$ we define the μth power of the fitness via $q_i = (F_i^{[k]})^\mu$.

(b) Let $Q_k(\mathcal{P}_k)$ denote a probability distribution over the m elements of $\mathcal{P}_k$ such that the probability of drawing $\mathbf{y}_i^{[k]}$ from $Q_k(\mathcal{P}_k)$ is exactly equal to $q_i / (\sum_{i=1}^{m} q_j)$.

(c) For $i = 1, \ldots, m$ we generate a new element of the next pool, $\mathbf{y}_i^{[k+1]} \in \mathcal{P}_{k+1}$, by first drawing parents $\mathbf{y}_a$ and $\mathbf{y}_b$ independently from $Q_k(\mathcal{P}_k)$ and then drawing $\mathbf{y}_i^{[k+1]}$ from $C(\mathbf{y}|\mathbf{y}_a, \mathbf{y}_b, X)$.

Clearly members of the kth pool with relatively higher Fs (fitnesses) stand relatively larger chances of being chosen as parents of the individuals in the next pool. The larger μ is, the more we force convergence towards the highest scoring member of the present pool acting as one or both parents of the offspring. We may think of the distribution Q_k as a roulette wheel for parenthood, biased by the q_i. So larger μ tends to force early convergence, while larger p tends to force the pools away towards the random picks that we had in generation one.

We iterate for as many generations as we require, remembering the best-ever solutions observed over the course of the evolution. Often in practice the GA is terminated when either a maximum number of generations has been produced, or a satisfactory fitness level has been reached for (some fraction of) the population.

In theory we could think of the situation where $m \to \infty$ so that the pool is really a present density distribution (proportional to fitness defined over Y). Then the operations of crossover and mutation can be applied to randomly chosen parents from that distribution, resulting is a new distribution over Y.

7.4 A Dynamic Model for Behavioural Changes

We consider customer journeys as discrete Markov chains [43], as introduced in Section 7.2. For a number of consecutive time periods, we shall describe each customer as belonging to one or another of a finite set of behavioural states (depending upon their behaviour within that time period). Then the period-to-period evolution of any customer will be described by a sequence (or chain) of states. Each individual chain evolves according to probabilities (it is thus a special class of random variable), and we can describe their future movements probabilistically. Moreover, we can aggregate over a population of customers and their corresponding chains. The resulting population represents a large sample (hopefully, if we have lots of customers) of the random variables (the chains) satisfying the Markov process.

Suppose all customers are divided into exactly n behavioural states at the end of each time period, based on their behaviour within that period. We shall let $\mathbf{x}_k$, an n-vector of real values, denote the expected distribution of customers over the set of n states at the end of time period k. We shall often include one of the states (state 1) called the 'lapsed' state, which represents those customers who did nothing, and made no contribution to value in a period.

In practice, the exact values historically for the distribution of customers over the set of n states, $\hat{\mathbf{x}}_k$ say, are known and contain integer counts, and we will use $\mathbf{x}_k$ to denote our expectation of this distribution in the future. We desire that the true value $\hat{\mathbf{x}}_k$ will remain close to its expected value $\mathbf{x}_k$ over time.

As before, assume there is a matrix A, called the *transition matrix*, such that A_{ij} represents that conditional probability

A_{ij} = Prob(In state j at end of next period | In state i at end of current period).

If there are no new customers, the expected distribution will evolve through successive time periods according to the equation

$$\mathbf{x}_{k+1} = A^T \mathbf{x}_k. \tag{7.1}$$

Here, A is a non-negative matrix so we might assume it is irreducible, but this is not necessary. For such matrices, the Perron–Frobenius theorem (see Chapter 1) states that the spectral radius $\rho(A)$ is itself an eigenvalue. It is bounded above by the maximum row sum and maximum column sum, and is bounded below by the minimum row sum and minimum column sum. Notice that in our application the ith row sum of A represents the probability that a customer in the ith state at the end of period k is still a customer (and in one or other state) at the end of period $k + 1$.

If the Markov modelling assumption is valid, then the actual distribution of customers at the end of the next period, $\hat{\mathbf{x}}_{k+1}$ will be very well approximated by the expectations $\mathbf{x}_{k+1}$. This requires

- the sampling errors to be low: we shall assume this is the case here, since we wish our model to apply to service providers like mobile telephone companies, or retail loyalty schemes, or online traders that have thousands or millions of live customers;

- the stationarity of the underlying process: there is a single transition process (governed by A) that is valid for all period-to-period transitions;
- the transition matrix, A, to be estimated as desired from available historical data.

We may include new customers acquired into states during period $k + 1$. Let $\mathbf{b}_{k+1}$ denote the distribution of such new customers during period $k + 1$, then our equation (7.1) becomes

$$\mathbf{x}_{k+1} = A^T \mathbf{x}_k + \mathbf{b}_{k+1}. \tag{7.2}$$

Later we shall be imposing some state definitions for the behavioural states, unambiguously allocating customers to states on the basis of each period's behaviour, and then calculating a suitable estimate for A as a consequence of the observed state changes over some historical data. In businesses such as retail we may decide that lapsed customers will stay within the lapsed state forever (in which case we lose no customers ever) and consequently the total number of customers, given by $\mathbf{s}^T . \mathbf{x}$, where $\mathbf{s} = (1, \ldots, 1)^T$, grows forever. Or we may take the pragmatic view that some customers in the lapsed state will leave forever, and in this case the corresponding row sum for A will be less than unity (somehow not all lapsed customers remain until the next period), and consequently the total number of customers will remain bounded (under reasonable assumptions on A and the $\mathbf{b}_k$).

Now if $\mathbf{b}_k = \mathbf{b}$, a constant vector for all k, and $\mathbf{x}_k$ approaches some steady state, $\mathbf{x}^*$, then

$$\mathbf{x}^* = (I - A^T)^{-1} \mathbf{b}.$$

Moreover, if the spectral radius of A is strictly less than one, then this iteration in (7.2) will converge to the steady state, $\mathbf{x}^*$ (by the contraction mapping theorem). On the other hand, if A is such that the row sums of A are all unity (that is that no customers ever leave the business), then $(I - A^T)$ is not invertible; equivalently, unity is an eigenvalue of A and hence the total number of customers, $\mathbf{s}^T \mathbf{x}$, grows forever.

Now consider *customer value*. Let $\mathbf{v} = (v_1, \ldots, v_n)^T$ (a real n vector) denote the vector of expected (mean) total revenue derived from a customer-period within each of the states. We can, and we will, estimate $\mathbf{v}$ from historical data, and indeed a customer's revenue over a period is likely to be a key attribute used in defining the states themselves (and hence in allocating customer-periods to the states). Then the expected total revenue for period k from all customers is

$$R_k = \mathbf{v}^T \mathbf{x}_k.$$

Suppose we are now at the end of period k. Then using (7.2), our expectation of total revenue from all customers from the next period is

$$R_{k+1} = \mathbf{v}^T \mathbf{x}_{k+1} = \mathbf{v}^T (A^T \mathbf{x}_k + \mathbf{b}_{k+1}).$$

Similarly our expectation of total revenue from the next but one period is simply

$$R_{k+2} = \mathbf{v}^T \mathbf{x}_{k+2} = \mathbf{v}^T (A^T (A^T \mathbf{x}_k + \mathbf{b}_{k+1}) + \mathbf{b}_{k+2}),$$

and so on.

If $\mathbf{b}_k = \mathbf{b}$ is assumed to be a constant, then

$$\mathbf{x}_{k+m} = (A^T)^m \mathbf{x}_k + \mathbf{b} + A^T \mathbf{b} + \cdots (A^T)^{m-1} \mathbf{b}.$$

These last terms might be written

$$\mathbf{b} + A^T \mathbf{b} + \cdots (A^T)^{m-1} \mathbf{b} = (A^T - I)^{-1}((A^T)^m - I)\mathbf{b}$$

when $(A - I)$ is invertible.

Now suppose that future revenues are discounted by a factor $(1 - \rho)$ for each period. Then the next period's total revenue is worth

$$\rho R_{k+1} = \rho \mathbf{v}^T \mathbf{x}_{k+1} = \rho \mathbf{v}^T (A^T \mathbf{x}_k + \mathbf{b}_{k+1}),$$

and hence we may similarly determine future revenue forecasts for subsequent periods, in absolute and discounted terms.

Next we turn to the value of an individual customer. What are our current customers worth now and in the future?

Consider a single customer within state i at the end of the current time period. His probabilistic state distribution is simply the ith unit vector, denoted by $\mathbf{e}_i$, say. At the end of the next r periods his probabilistic state distribution is $(A^T)^r \mathbf{e}_i$.

This customer's value in state i during the current period is given by $\mathbf{v}^T \mathbf{e}_i$. His value in the next period is simply the discounted expected revenue based on our estimation as to which state he will be in: $\rho \mathbf{v}^T A^T \mathbf{e}$. His value over the next r periods is simply the sum of discounted expected revenues:

$$\rho \mathbf{v}^T A^T \mathbf{e}_i + \rho^2 \mathbf{v}^T (A^T)^2 \mathbf{e}_i + \cdots + \rho^r \mathbf{v}^T (A^T)^r \mathbf{e}_i = \mathbf{v}^T (\rho A^T + \rho^2 (A^T)^2 + \cdots + \rho^r (A^T)^r) \mathbf{e}_i.$$

We define this to be his *Customer Horizon Value*, denoted by HV_i: the discounted revenue or the net worth of the future revenue over the next r periods, derived from a customer currently in state i. We shall refer to the next r periods as the horizon period.

If $\rho < 1$, or if the spectral radius of A is less than unity, then $(I - \rho A^T)$ is invertible. Hence

$$HV_i = \mathbf{v}^T (I - \rho A^T)^{-1} (\rho A^T - \rho^{r+1} (A^T)^{r+1}) \mathbf{e}_i.$$

Furthermore, we can let r tend to infinity and talk about the *Customer Lifetime Value*, CLV_i, for customers currently in the ith state:

$$CLV_i = \mathbf{v}^T (I - \rho A^T)^{-1} (\rho A^T) \mathbf{e}_i.$$

7.5 Defining Models

There are broadly three ways of defining states that may present themselves within any business.

Using an Existing Customer Segmentation

Many companies use their existing customer database for marketing purposes, and often specifically for direct marketing by themselves and their suppliers or partners (mail outs, vouchers, emails, invitations, etc.). Very often they will use a *segmentation* of their customers. This is simply a partition based on inferred or actual lifestyle information, on geographical information, or on long-term behaviour. For example, there may be a segment of family buyers, or empty nesters, or low value-low loyalists, young couples, single parents, or retired homeowners, and so on. Such attributions are usually made once a year, or once every two years. Assuming they represent a proper partition, the segments may be used as states provided that we can select a suitable definition of the time period so that we can redesignate customers to segments (now called states) regularly. This may be problematic. In order to enter some of the states you may have to get married, have a baby, lose your job, retire, or move house. What we must accept is that the world has changed and people's behaviour evolves far faster than their circumstances. For some segments there is a wide variety of possible behaviour, for example within a segment like seniors (OAPs) there are some members behaving like forty somethings, and there are other members who hardly ever go out at all.

If the states have such definitions that would take individuals years to migrate between, then this is not going to be useful for our purposes. It is a snap shot, not a dynamic model.

Simple Behavioural State Definitions

Choose a time period that represents the time scale on which you can market and affect customers, or over which you wish to forecast and review (say a month or a quarter). How often do customers transact within that period? Extend the period to make sure that most customers who trade do so two or more times. Now select two or three measures that reflect customer behaviour, and value. For example, total value over the period (value), and number of transactions within the period (frequency). Sometimes value-frequency-recency is used. Recency could be measured by the number of days from the last transaction up to the period end.

Partition this two or three dimensional space into states with a grid, or by simple clustering (as in Chapter 4).

Discovering Complex Behavioural State Definitions

In this method, states are to be defined in terms of a (possibly large) number of behavioural attributes summarizing the customer's behaviour within a time period (up to twenty or so to keep the computations manageable). Conceptually, though, the idea is similar to the last method. As before, the time period must be defined to be long enough so that there are at least a few transactions on average within each period. Typically a period may be a few weeks or a month or even a quarters as described above.

It is desirable to choose the state definitions so that they yield a good description of the full range of behavioural types exhibited across the customer base, and so that the period-to-period customer movements are dominated by a few key transitions.

The second requirement is not as obviously necessary as the first. Clearly if the states' definitions, based on current behaviour, were no indicator of the future behaviour of the customers then there would be similar flows from all states into any particular state (all the rows of A would be similar), and such a model would have no predictive power at all. Hence, in choosing the model's set of state definitions, it is necessary to consider state-to-state transition rates in the corresponding transition matrix. By optimizing the definitions of the states so as to obtain a model dominated by a few transitions and making robust estimates of these transition rates, the customers currently in distinct states would possess distinct futures (or distinct expectations of their future behaviour and hence value). The result would be a model of the evolving (dynamic) behaviour changes of customers, which will be used to present expectations about their future behaviour.

The second requirement is equivalent to making the transition matrix sparse and robust to sampling errors. This will also mean that the users (the marketing team) will be focussed on a few important transitions that drive or lose value to the business, and the transition rates in the model can yield useful predictions.

How should the states be calibrated and optimized? It can be achieved efficiently by using a genetic algorithm to define and score many possible models [48]. We have used GAs for the case studies presented later. A suitable fitness function should reward both the sparsity for A and a measure of the relative uniformity of the state populations. This optimization may take a number of hours to achieve, but it only needs to be done once.

Whatever method is chosen to define states and transitions, the analyst will almost certainly be challenged on the applicability of the model and the meaning or characteristics of the states. Traditionalist marketing executives often have a curious attachment to the simplest and suboptimal models that are nevertheless explainable in terms of a few key attributes. So the more complex the analytics, the more trust will be required, and this is only earned by the returns on investments realized by the results.

Once we have settled our definitions for the states we will need to use historical data in order to produce an estimate for A. Then suppose there are exactly r_i customer-time periods within state i for which we also know the customer's subsequent state (in the next time period). Let this number also include those customer-time periods where the customer subsequently left the business (closed the account) in the next time period, if any such exist. Then suppose exactly M_{ij} of these move to state j in the subsequent time period. Note if there are no closures we will have

$$r_i = \sum_{j=1,\dots,n} M_{i,j}.$$

Then we use the estimate furnished by *Laplace's Law of Succession*, see the Appendix:

$$A_{ij} = \frac{M_{i,=j} + 1}{r_i + n},$$ (7.3)

valid for $j = 1, \ldots, n$ and $i = 1, \ldots, n$.

Note that the result can never be zero or unity: it lies between $1/M_i$ and $1 - 1/M_i$.

Exercise 7.3

Suppose we are given a sequence of state classifications as a sequence of integers $\{c_k | k = 1, \ldots, N+1\}$, where $c_k \in \{1, \ldots, n\}$, which we assume has been generated by a corresponding n-state Markov model. Let M denote the $n \times n$ matrix of counts for the transitions :

$$M_{ij} = \text{count of occasions when} c_k = i \text{ and } c_{k+1} = j.$$

Let $r_i = \sum_{j=1}^{n} M_{ij}$, as before, denote the row sums of M. Then our estimates for the elements of the transition matrix, A, using Laplace's law of succession, are given by (7.3).

Assuming that the individual state transitions are independents show that the likelihood of the data given the model, $\mathcal{L} = P(\text{Data}|\text{Model}, X)$, satisfies

$$\log \mathcal{L} = \sum_{i=1, j=1}^{n} M_{ij} \log (1 + M_{ij}) - \sum_{i=1}^{n} r_i \log (n + r_i).$$

7.6 Seasonality: Manageable Non-Stationarity

One of the basic assumptions made in employing a Markov model is that the customer transition process was the same from each period to the next period. So in that case A can always be employed to make a good prediction of next period's population distribution based on the current one. Suppose this is not so, because our time period is too short (chosen for marketing purposes) or there is some seasonality within the model. For example, Christmas and holidays often produce seasonal changes in customer behaviour. The framework we have described may be modified easily to deal with this case.

Suppose that the year is divided exactly into K periods, say $K = 13$ four-week periods, or $K = 4$ quarters. Then we will calibrate K transition matrices A_k to model the transition from the kth period of a calendar year into the $(k + 1)$th period. We need data from at least two years back to do this, and we only use the relevant transitions to calibrate each transition matrix. If all of the A_k's turn out to be the same then we need not worry about stationarity. However some of them may differ, indicating certain seasonal behaviour changes. In that case we can forecast just as we did before but we have to be careful to use the correct estimate for each corresponding A_k as we predict populations forwards from each kth period of the year to the next period. This is just nasty detail. Population transition matrices, forecasts, horizon values, and CLVs can all still be calculated. We can also predict what state folk will be in at say Christmas (period K). If $\mathbf{x}_j$ denotes our current population at the end of period j of the current year, then

$$ A_{K-1}^T A_{K-2}^T \ldots A_{j+1}^T A_j^T \mathbf{x}_j $$

represents our forecast for these customers' behaviour at next Christmas.

7.7 Case Studies

We present some anonymous case studies so that practitioners may have some comparisons with their own customers' transitions, values, and lapse rates, since rarely will such information be shared.

A Large Grocery Chain

Here is a fifteen-state model based on more than eighteen month's customer account card data. Transactions were split into monthly time periods with each customer-month represented by thirteen input attributes. These included information about:

- the volume/value of transactions
- the types of transaction (content etc)
- inferred estimates of other spending
- take-up of promotions
- distinct number of stores and formats visited.

Table 7.1 *Customer Monthly States*

Name	State i	% of live customers	v_i, Period Value ($)	% Rev Contrib'n
Low Value Low Freq	3	21.2	57.70	5.1
Low Value Medium Freq	6	5.8	64.45	1.6
One Main per Month	2	9.4	99.99	3.9
Three Shops Weekly	11	2.1	121.69	1.1
Two Means Four TopUps	9	6.1	122.84	3.1
Irregular Two+ Stores	7	14.1	189.62	11.2
Every Other Day	13	3.1	222.92	2.9
Three Mains per Month	4	9.6	253.43	10.1
High Value Med Freq	12	14.7	248.00	28.1
High Value High Freq	14	1.7	459.17	3.3
V High Value Low Freq	5	5.9	473.04	11.6
V High Value Med Freq	8	3.0	484.33	6.0
Constant Shoppers	15	0.4	658.04	1.0
Super Value	10	3.1	866.83	11.1

Table 7.1 details all 14 live (trading) customer states in value order together with data for each state, taken over 0.25 million customer accounts.

Lapsed customers are around 20% of the total customer accounts: 'live customers' refers to the 80% of the total who are trading each month. Hence 3.1% of the non-lapsed customers are in the Super Value state, contributing 11.1% of the retailer's total revenue. Slightly more than 61% of the total revenue comes from the 28% of non-lapsed customers in the six most valuable states (21% of all customers).

The transition matrix is given next, in Table 7.2, representing the probability of moving from the row state in the current month to the column state in the next month. The states are not ordered here by period spend (see Table 7.1), but are in the order in which they were discovered and named. The matrix is sparse because we have optimized the state definitions for this to be so. We did this for each of these case studies using a genetic algorithm to find centroids and weights as described above in Section 7.3.

Table 7.2 Month to Month Transition Matrix

Name	State	1	2	3	4	5	6	7	8	9	10	11	12	13	14	15
Lapsed	1	.61	.10	.17	.02	.02	.01	.05	0	.01	0	0	.01	0	0	0
One Main per Month	2	.23	.21	.19	.12	.06	.02	.13	.01	.01	.01	0	.02	0	0	0
Low Value Low Freq	3	.18	.08	.34	.05	.01	.09	.12	0	.07	0	.01	.04	0	0	0
Three Mains per Month	4	.03	.08	.08	.39	.06	.01	.15	.06	.03	.01	0	.08	0	0	0
V High Value Low Freq	5	.06	.07	.03	.09	.32	0	.10	.08	0	.15	0	.09	0	0	0
Low Value Medium Freq	6	.04	.02	.24	.02	0	.35	.8	0	.18	0	.04	.02	.01	0	0
Irregular Two+ Stores	7	.06	.07	.14	.10	.04	.04	.32	.02	.07	.01	0	.12	.01	0	0
V High Value Med Freq	8	0	.03	.02	.18	.13	0	.09	.31	0	.09	0	.16	0	0	0
Two Means Four TopUps	9	.01	.01	.14	.04	0	.13	.12	0	.34	0	.07	.08	.04	0	0
Super Value	10	0	.01	0	.03	.22	0	.03	.07	0	.49	0	.13	0	0	0
Three Shops Weekly	11	.01	0	.05	0	0	.10	.02	0	.19	0	.38	.09	.16	.01	0
High Value Med Freq	12	0	.01	.03	.05	.03	.01	.10	.03	.04	.03	.01	.60	.04	.02	0
Every Other Day	13	.01	0	.02	0	0	.02	.03	0	.09	0	.10	.17	.45	.11	0
High Value High Freq	14	0	0	.01	0	0	0	0	0	0	0	.01	.18	.20	.52	.07
Constant Shoppers	15	0	0	0	0	0	0	0	0	0	0	0	.02	.03	.28	.66

Look at the monthly lapsing rates in the first column. This is typical for large stores in both the US and the EU that are selling grocery and non-food. There is highest consistency among the Constant Shoppers: these people come in every day, about thirty times per month.

The very first time that the marketing and business folk within a retailer see this data (and the consequent value, horizon value, and forecasts), they will almost always be fascinated and surprised as to how volatile the customer base appears. They tend to see everything summed up so it is close to a monthly dynamic equilibrium. This means there are opportunities. Often users will overlay the states with a static lifestyle, or social segmentations if available. Though they may be familiar with some of the major effects, this is the first time for many that they can see hard numbers for the transition rates. In our experience it is critical for the analyst to explain the state definitions and the model in simple terms, and to walk potential business users through these summary results. This gives them confidence and gets buy-in.

Customer value can be improved by increasing the frequency with which customers shop. For a similar large grocery retailer in the EU we identified a medium value state and high value state that were defined by behaviour that differed only by frequency (they were monthly balanced shoppers and weekly balanced shoppers), rather than by basket sizes or contents. The transition matrix provided hard evidence that there was existing switching between these two states (around 10% to 20% of both states), so they were indeed the same people. Moving as many as possible of these up to the weekly balanced shopping state increased their monthly value by a factor of four. This was a clear marketing task and it had a clear return on investment.

There was an alternative opportunity within the same model. We identified two other states where customers were buying the same types of items but in smaller and larger baskets, respectively (smaller and larger total value). There was a considerable month-on-month exchange of customers between those two states, so again it could not be argued that they were specifically small and large basket shoppers. This repeated switching pointed to low loyalty. Holding such customers up in the higher value state, or moving more up there, increased their monthly value by a factor of two. Messaging was be built around 'basket building' within the key product ranges for such customers. Note basket building is quite distinct from messages that are designed to increase frequency, but let us allow the marketing experts take care of that.

An Online Grocery

We wanted to estimate the value of the growing customer base of an online retailer based on their account transaction data.

Customer data used was as follows. We used six consecutive periods each of six weeks of customers' collected transactions (six weeks in order to allow for a reasonable amount of repeated purchasing). We measured value ($) of the online orders aggregated into 'macro categories':

- Personal care
- House, garden, and pet
- Alcohol and tobacco
- Ambient food items
- Fresh food items.

Customers could lapse and return; within each period we also calculated:

- spend volatility calculated over fortnights within the six-week periods
- weekend spend skew
- evening spend skew.

The suggested model had ten live states and one lapsed state, see Table 7.3. The mean period spend for each of the live states varied between £96 or so up to £496 and £635 for the two most valuable states. The mean number of transactions within each six-week period for customers in each state varied between one and five (almost one transaction per week in the two most valuable states). There was a cluster of five states spending on average less than £130 and making 1 to 1.6 transactions per month. The states in the cluster were differentiated through their behaviour rather than just the customer value and frequency.

New accounts were acquired into all states and accounts could transfer from state to state. Accounts could lapse, but also return from a lapsed state.

Note that customers in 'home and pet' and 'Work-time BWS' (BWS stands for 'beers, wines and, spirits') were less valuable over the horizon period than we would expect compared to the other states.

The transition matrix given in Table 7.4, represented the probability of moving from the row state in the current period to the column state in the next period. The states were ordered by period spend. Note that lapse rates were not decreasing with period spend. In general they were much higher than those seen in bricks-and-mortar businesses. This matrix is sparse because it

Table 7.3 *Customer-6 Week States*

Name	State *i*	Pop'n (%)	v_i Mean Value (£)	Evening Skew (%)	Weekend Skew (%)	HV (£) (4 periods)
Lapsed	1	40	0	0	0	199
Fresh food	2	1	96	0	16	221
Work-time BWS	3	1	111	0	0	198
Work-time food	4	19	121	0	0	333
Weekenders	5	11	125	0	97	347
Personal care	6	5	128	0	9	368
Evening	7	3	180	94	11	409
Home and pet	8	1	213	0	23	256
Balanced	9	11	237	5	55	478
High value	10	7	496	4	34	871
Very high value	11	3	635	2	34	862

was optimized by a GA. Customer value growth was driven by just a few transitions in behaviour; see the transition matrix in Table 7.4.

In the US there was a very famous example of a pure-play online grocery chain that went wrong and crashed. Webvan had an enormous valuation when it went public during the dot-com boom. After its demise we analysed their customer data. Webvan's problems stemmed from a simple inability to hold onto, and grow, its most valuable customers. The customer data showed huge lapsing rates, yet the cost of acquisition was very high. The rapid and expensive early growth of the customer base was supported by a costly and over-optimistic delivery infrastructure, but the churn rates meant that the customer density did not stay high, and as a result the cost of delivery (the last few miles) did not scale.

A Mobile Phone Company

Here we present a model based on mobile phone usage data calibrated over 500,000 contract accounts. The most satisfactory model (after optimization by a GA) had twelve behavioural states.

Table 7.4 Transition matrix

Name	State	1	2	3	4	5	6	7	8	9	10	11
Lapsed	1	.83	.0	0	.07	.04	.02	.01	0	.03	.01	0
Fresh food	2	.62	.05	0	.33	0	0	0	0	0	0	0
Work-time BWS	3	.70	0	.10	.05	.10	.05	0	0	0	0	0
Work-time food	4	.51	0	0	.23	.07	.03	.02	0	.08	.04	.02
Weekenders	5	.51	0	0	.14	.18	.02	.01	0	.08	.04	.03
Personal care	6	.47	0	0	.15	.05	.13	0	0	.10	.09	0
Evening	7	.38	0	0	.08	.03	.01	.32	0	.14	.05	0
Home and pet	8	.53	0	0	.13	0	0	0	.33	0	0	0
Balanced	9	.34	0	.01	.13	.07	.05	.03	0	.24	.09	.04
High value	10	.10	0	0	.08	.04	.07	.02	0	.15	.47	.07
Very high value	11	.11	0	0	.17	.10	0	.01	0	.10	.19	.32

Each month the customers were defined by a set of attributes including those summarizing:

- total usage by value
- roaming by value
- international calling by value
- SMS by value
- UK calls by value
- data and fax by value
- total number of calls
- number of calls made to fixed lines and other networks.

Twelve behavioural state were identified. A summary of the number of customers which initially belonged to each of these states is shown in Table 7.5, along with other derived measures. The time period selected was monthly, as this corresponded with the billing period. We did not impose a lapsed state, although we did have a 'tiny' usage state that included lapsed customer-months.

To calculate the horizon values we took $\rho = 1$ (no discounting). The transition matrix for this model, A, was given by:

Table 7.5 Mobile Phone Customer-Monthly States

Name	State	Pop'n (%)	Usage Value (£)	UK Voice Skew (%)	No in/out Calls	HV (£) (6 months)
Tiny	1	2	1	68	99	89
Low value voice	2	5	2	86	80	82
Low value roamers	3	27	4	35	12	56
Voice (short) and SMS	4	10	4	82	42	74
Voice and SMS	5	17	6	78	24	62
Low value roamers	6	6	8	54	31	85
Voice and data users	7	1	11	28	371	60
Med value voice	8	8	16	92	131	115
Med value roamers	9	5	19	62	151	139
Heavy voice and SMS	10	7	30	59	173	113
Voice SMS, int'l	11	6	39	45	257	162
High value roamers	12	4	90	55	259	215

$$\begin{pmatrix}
0.016 & 0.235 & 0.124 & 0.174 & 0.029 & 0 & 0 & 0.42 & 0.001 & 0.001 & 0.001 & 0 \\
0.001 & 0.242 & 0.162 & 0.062 & 0.205 & 0.005 & 0 & 0.058 & 0.001 & 0.242 & 0.021 & 0.001 \\
0.001 & 0.019 & 0.71 & 0.037 & 0.141 & 0.026 & 0 & 0.026 & 0.005 & 0.024 & 0.007 & 0.003 \\
0 & 0.03 & 0.113 & 0.356 & 0.246 & 0.079 & 0 & 0.069 & 0.019 & 0.057 & 0.025 & 0.006 \\
0.001 & 0.047 & 0.268 & 0.129 & 0.453 & 0.014 & 0 & 0.02 & 0 & 0.061 & 0.006 & 0.001 \\
0 & 0.001 & 0.129 & 0.123 & 0.044 & 0.525 & 0 & 0.044 & 0.074 & 0.006 & 0.032 & 0.022 \\
0 & 0 & 0.03 & 0.01 & 0.059 & 0 & 0.9 & 0 & 0 & 0 & 0 & 0 \\
0 & 0.017 & 0.088 & 0.068 & 0.052 & 0.036 & 0 & 0.465 & 0.064 & 0.104 & 0.045 & 0.059 \\
0 & 0.001 & 0.034 & 0.034 & 0.004 & 0.101 & 0 & 0.11 & 0.574 & 0 & 0.04 & 0.101 \\
0.001 & 0.026 & 0.07 & 0.023 & 0.134 & 0.001 & 0 & 0.052 & 0.001 & 0.638 & 0.041 & 0.015 \\
0 & 0.015 & 0.037 & 0.039 & 0.02 & 0.036 & 0 & 0.057 & 0.029 & 0.068 & 0.625 & 0.075 \\
0 & 0.001 & 0.039 & 0.015 & 0.007 & 0.023 & 0.001 & 0.105 & 0.094 & 0.035 & 0.113 & 0.567
\end{pmatrix}$$

Consider the most valuable customers within state twelve, The high value roamers state. Only a fraction of 0.567 of them stayed where they were from month-to-month. This business lost more revenue from its existing customers modifying their behaviour into successively lower value states (consciously or not) than through lapsing.

A Personal View

On Non-Open Methods and Data

The proprietary nature of customer data coupled with the popular strategic objectives of defending and growing value from current customers means that most organizations that have customer-centred data must keep them it carefully locked away. Indeed, it is really the most precious asset that an incumbent player within their sector has. How can we square this with a wish to have an open ecosystem of research and technology development? Most often companies will have an internal consumer insights analytics group, or something similarly named that is not open to outside participants or benchmarked properly. Hence neither the data nor the methodologies employed there are open. I would argue that this is mostly a necessary consequence of competition. Worse, if sector analysts get hold of raw customer data they might make unfavourable comparisons between the performance of a company against its competitors. This would have a dire impact on share price. So the data are not just customer sensitive, their unconstrained release would be a risk to shareholder value.

Sixteen years ago, together with others, I founded Numbercraft and began working on algorithms for retailers and consumer goods companies. We always respected the confidentiality requirements of each of our large data-holding clients, and while we could work with those clients in an intimate way, we could rarely (if ever) publish anything at all. This suited us too, since it was impossible to get any legal protections for our algorithms, and our own investors' interests would not be well served if all of our research and discoveries could be exploited immediately by our competitors for free. Without patents we were forced to protect our know-how through secrecy. We also had to manage considerable competition between our clients, so we could never leak ideas or even their expressions of interest in certain topics. At the time it seemed strange to me (though it is obvious in hindsight) that the more secretive we were the more trusting our clients became, and the more enhanced was our reputation. It got to the point where our clients, such as PepsiCo, would present selected results of our work and its impact at conferences and more privately to retailers, referencing Numbercraft, in a manner that we did not. We adopted a general 'I couldn't possibly comment' stance. While we retained the ownership of all of our concepts and methods, the data and actionable insights within any engagements belonged to the client.

Recently there have been two developments that may make things easier for academic researchers and start-ups to test out radical ideas on real data.

One is the emergence of competition sites such as Kaggle (https://www.kaggle.com). Here, competitions and data are made available by owners (in some suitably anonymous fashion), and registered competitors can compete with one another according to the declared performance measures. Counting Lab, an analytics technology translation company founded in Reading, entered a Kaggle competition on forecasting energy demand. They not only gained contacts with other groups broaching similar problems in very different ways, but they benchmarked their own approach, attended an international workshop for the top few competitors, and gained some exposure and publicity. For the data owners there is real value here, since they can leverage their effort in setting up a relatively small amount of anonymized data into gaining effort from a possibly very large number of individuals, academic groups, and small company teams the world over. This is a true win-win. I encourage you to check out Kaggle and similar sites if you have not done so.

The other development is the growth of open data initiatives designed to encourage innovation in various countries, including the US and the UK, where public sector data sets are released under a 'creative commons' compatible license. This enables the development of novel applications and services, especially by entrepreneurs and start-ups that would otherwise have been unable to demonstrate their ideas and performance. Primarily this has produced open access data on environment, weather, transport, maps, households, crime, geneology,

continued

On Non-Open Methods and Data *Continued*

and other publicly funded and collected data sets. Academic research is also included, where data sharing is nowadays usually a condition of funding when university-based projects are wholly funded from the public purse. In the UK this means that data sets cannot remain closed once a major publication has relied on it. Note that it does not allow the researchers to retain sole use of their data until *all* of their possible publications have been made. Such rules prevent the taxpayer having to pay for the repeated collection of similar data sets on separate projects, and it also encourages larger, multi-contributed data collections. However some mystery shopping—calling up the grant holders—indicates that at best there is a lack of awareness of this commitment, and at worst a wilful disregard of it. The true picture is mostly the middle. As data sets get larger it is actually becoming very difficult and costly to publish data and provide long-term curation and a quality level of service to future exploiters.

Additionally, we are seeing governments setting up customer-empowering data initiatives where those companies that hold customer and consumer data allow individuals to access their own data so as to gain insights into their own behaviour, make more informed choices about products and services, and manage their lives more efficiently. Of course the problem with this is that it undermines the commercial advantage to those presently with the private loyalty card and account cards. It might open the door to new companies who could create their own large-scale longitudinal data by aggregating the data from lots of individuals (by acting as their agents). So most likely the companies that comply with such an schemes will be those that are in sectors that have strong regulators (for example, energy, finance), and that do not currently exploit data for competitive customer insights. It is hard to see retailers or e-commerce players parting with their hard-earned data, which they have usually traded benefits to collect and are often run on an opt-in basis.

However this unfolds, we are today living in an era of ever-increasing consumer data that is not anonymous, and that can be linked to individuals, either directly or with a little bit of work. This is also an era of public empowerment. So those who would analyse and exploit the citizens' data must be very clear about the benefits to the individuals. Indeed retailers have always been clear about this. To not do so will result in mass push-back.

On the Qualification of Prospects and Topic-Led Conversations

Only very rarely would any potential exploiter of analytics say 'can we explore the use of concept X or method Y?' or 'I hear you have an exciting way of dealing with real-time data streams, let's test it out.' Almost all companies that are customer-facing and need to employ analytics will be driven by topics, areas of challenge, and the value of any potential solution to the problems they pose. For this reason, mathematicians working in analytics need to develop some empathy with their users' viewpoint, and be able to see both the utility and limitations of particular ideas and methods in analytics from that perspective. Indeed, once you have a good relationship with your exploiters it is always possible to open conversation around their topics. 'I've been doing some thinking about increasing loyalty and value of occasional customers. Can I come and talk to you about this?' 'What is new in pricing and promotions? I have some ideas for you.' 'Are you interested in becoming more strategic about your use of social media?' Most companies have a road-map or some kind of priority list of topics to resolve. It changes every quarter in response to internal and external events, results, impacts, and opportunities. The topics are themselves often cyclic and, as the technology changes, the customer evolves, or new competitors emerge, they come around in turn.

Many conversations with commercial entities who are prospective partners or co-funders of our research or prospective clients of our company are actually just a waste of our time. However passionately we may wish to work with them (to extend our methods and our applications, or to exploit and achieve some impact) we have to expect that there are many hard reasons for them not to act on the other side. The two biggest stoppers to any plan to work together, especially in innovative analytics, are 'not invented here' syndrome, where the prospect really only wishes to do things themselves in some internal group, and 'do nothing'. The latter is usually accompanied by a desire that something easier or less expensive will turn up that might be less risky or more effective. For this reason, whether we are in an academic group or a small company, one must always qualify one's prospects hard.

The most valuable thing that we have is our time. We should not spend more of it on those prospects that will not perform for us, so I often advise people working in business development, knowledge exchange, or innovation to be clear and open about their aspirations inside meetings with prospective partners, and qualify them according to a number of unconscious signals that they send back out to you. Note that your own outreach staff or business development staff will not do this, since your time, as a technical person, is free to them and they will be willing to give any prospect the kiss of life, as necessary. They always want the next meeting in the diary. This topic really requires another book, but my central point is that a proper method of dealing with this can be learned, and we analysts should not in general sacrifice rationality in the emotional desire to make something happen.

We should have a framework. If a prospective partner does not ask about who else you are talking to, what the costs of doing something would be, why

continued

On the Qualification of Prospects and Topic-Led Conversations *Continued*

the costs are so high, or who would own the results; if they cannot explain their decision-making process, their internal sign-off, or their data readiness; if there is not an obvious champion on their side, or they do not spontaneously espouse something you believe to be a driver for the collaboration; if any or all of these warning flags pop up then it probably is not going to happen. Even if they do some of these things, and have a corporate mission to widen their innovation ecosystems and interact with more people such as ourselves, we should ask ourselves whether this is just about them having an actionable option, to be drawn down if ever needed, or whether they will actually perform.

As soon as you decide that the next meeting is not worth it, you will save a extra day to do some more analytics or to see somebody else. Remember that both sides are qualifying each other, and you have a vote and a veto on whether to proceed.

I am keen that analytics professionals should defend their time. There are only so many golden hours in the day when one can get into the zone and engage theoretically with complex problems. We need to become adept at making the most of these hours, and then use the rest of our time effectively: to be involved within business development and knowledge exchange, to write things up, or to do some coding and numerical testing.

APPENDIX: UNCERTAINTY, PROBABILITY AND REASONING

Uncertainty and Making Decisions

In many situations, whether in business, society, the environment, the clinic, or the laboratory, we are able to monitor a wider and wider variety of performance parameters and events. For effective knowledge distillation, interpretation, prediction, and proactive intervention we need to develop some methods and algorithms. When we look at a complex or complicated system and make observations, what do we learn? What do the observations tell us about the object of our attention? The pace of change of many technologies means that concepts and methods should ideally be adaptive to new inputs, should be self-learning or self-tuning, and hence may be applied even as the underlying nature of the activity and the process of observation evolve. Analysts must deal with evolving uncertainties and express their estimation of these, and incorporate reasoning to make their understanding clear: they need methods that are genuinely much smarter.

By 'smarter' we do not mean an ability for making algorithms (or software) easy to set up or to integrate across data sources (that is called deployability), or the ability to provide and transmit content and support for such alerts. All of this is just efficient generation and use of 'intelligence', used in the sense of 'military intelligence' (the well known oxymoron). What we mean by 'smarter' is the sense that the algorithms can reason in a way that is consistent with the way that we ourselves think; that the algorithms are logical, and extend logic to common sense inferential rules; that the algorithms are used to consistently and continuously update the current view as to what is (or is not) certain. Moreover, a one-time decision (to change, to invest, to treat, to investigate, to interrupt, or terminate) may need to be taken for management reasons at some particular moment in time. In this case, best estimates of an evolving situation are required (What will happen under alternative decision scenarios? What is happening right now?). There is no time for the analysts to say, 'There is not enough data' or 'We do not know whether something aberrant can happen—no danger thresholds have yet been exceeded'. Instead, decision-making requires the support of our best current recommendations, even if there is little data. Designing algorithms to give good decision support requires us to draw the best possible inferences we can, based on everything we know to date (our prior knowledge), and our current incoming information (observations, data) about events.

What may be surprising is that properly viewed, probability theory is really just an extension of logic. It is not simply about making estimates of chance events, but it is far more powerful. Rather than asking, 'We know how the system works and what its uncertainties are: what is the chance of it producing an output like X?'; we want to ask 'Having observed an output like X, what does that tell us further about how the system works, and what its uncertainties are?'

A moment's thought and you will see that almost all plausible reasoning (from common sense to deduction) requires us to grapple with the second type of question. In business or in science this is ever more true, as we are able to capture or monitor more things. We observe events and results, and then we ask how does that alters our view about what is happening.

A good introduction to plausible reasoning is a good grounding for any decision-making. Careful use of language, inferences, and knowledge of uncertainties flushes out what we do not know (where our uncertainties reside and derive from), and points us to the information that we need to make better justified decisions. The literature abounds with apparent paradoxes—these usually come from poor application of just a few simple principles (legal, medical, engineering, commercial, etc.)—or from using the wrong conditional models to update our world view.

Hence an understanding of the material in this Appendix represents a skill for life.

Sources of Uncertainty

Uncertainties arise in different ways. There is the uncertainty over the truth of many propositions about the underlying system that we are addressing, and over the events that we are observing. The car is safe. The witness is lying. The economy will improve. The climate is warming. England will win the World Cup. What do our observations tell us about the system, and the truth of these statements?

Observations of any related events may change the uncertainty we have in asserting these propositions. To see this, we need to model the consequences of the proposition and alternative propositions, and see whether the observations are more consistent with one hypothesis or another.

Modellers, and especially mathematical modellers, are often uncertain as to which models they may apply to a given situation, though. They must make some decisions, some assumptions, and some omissions, based on their experience and sometimes on pragmatics. They must choose a class of model, and not just go to what they know, like men with hammers looking for nails. How can modellers compare one model against another? This is called *conceptual model uncertainty*. Every assumption modellers make in setting out a model introduces uncertainty, and models which share assumptions are not independent: if a common underlying assumption becomes invalid then so do the models. When we observe some events that can or cannot be represented well by the behaviour of solutions to a model, we become more or less convinced of its usefulness. Our uncertainty in the conclusions or predictions from the outputs decreases or increases accordingly.

Given a particular model it may well have parameters that we do not know, or that we are uncertain of, yet we have some experience or ideas about their value. This is called *parameter uncertainty*. We need to use evidence or observation to make better estimates for the possible whereabouts of parameter values.

Our models may contain variable or stochastic, space- or time-dependent, processes that are parameterized. This is called *volatility* or *variability*. You could think of these quantities as sophisticated parameters if you wish. These things need to be modelled too. This introduces submodels, with submodel conceptual and parameter uncertainties.

So, even armed with our prior assumptions and beliefs, our models, and our methods of making them operational (solution methods, numerical calculations), we may well need to actually find some things out. These may be inferences, forecasts, predictions, or may be decisions based on our estimation of uncertain model outputs. The more that we know or observe, the more we hope to evolve our representation of the uncertainties concerning such outputs.

Sometimes we have other challenges. For example, in watching an animal, a person, or a machine, and trying to understand the nature of (and constraints upon) its ability to reason with uncertainty and respond to its own perceptions or observations. We might be building or hypothesizing systems that could reason with information or possible facts. In these models we have to represent a process of reasoning, where the currency itself (the state variable) is uncertainty and is represented over 'possibility space'.

To face any or all of these challenges we should at first be very clear about the mathematics of uncertainty and reason.

Bayesian Probability

What a strange thing probability is. This point was almost entirely missed at school, where we were all taught an experimental view, with definitions derived from expectations: the probability of picking a heart from a standard pack of cards—try it out lots of times. We were encouraged to imagine doing the experiment over and over again. Then the probability was simply the number of successful events divided by the number of attempts. But this approach is only ever valid in the limit of an infinite number of experiments: we do not have the time or the imagination in complex cases. And what about situations where we cannot actually make or imagine repeated experiments because the question in mind is too unique, or we have only a very small amount of evidence?

In such cases, 'probability' still means something to us though: it can obviously represent our uncertainty surrounding the truth of a given proposition. A proposition is just a statement about some event which may happen, or has happened, or a statement about certain circumstances being correct:

> 'The card I select will be a heart',
> 'It is raining', ' I am a liar',
> 'The next president of the US will be a woman',
> 'Ghosts exist', or
> 'Professor G is a murderer'.

Look at these examples: the more interesting, controversial, or salacious a proposition, the harder it is to imagine any thought experiments that could determine an agreeable absolute probability which will satisfy all onlookers. The probability that such a proposition is true is entirely subjective (I may know more about the subject or have more information that you), so our opinions, and our assessments of the residual uncertainties, are personal to us. Subjective probability is really the currency of logical reasoning. Even though our estimates for the probabilities may differ between us in absolute values, whenever any new evidence arrives both of our subjective probabilities should be updated in a consistent manner. That is, if we share the same starting point (so that our prior estimates are the same) then the new information and evidence should alter our estimates in the same way, so that we arrive at the same 'posterior' estimate of uncertainty after we have accounted for the new evidence. This consistent updating process is called Bayesian Updating, named after the Reverend Thomas Bayes.

A Crash Course in Plausible Reasoning

So let us start probability theory all over again. Let us leave aside what we have learned up to now. This is all we will ever need and we shall hold to these central points.

Resources: three excellent and invaluable books to dip into are Jaynes and Bretthorst (2003), D'Agostini (2003), and O'Hagan and Forster (2004) [83, 32, 109].

1 *Plausibility.* For any statement, called a proposition, we refer to the 'plausibility' of the proposition meaning our belief that it is true. Sometimes we will have alternative, mutually exclusive, propositions, usually called 'hypotheses', where we are certain that one of them is true, and yet we are uncertain as to which one. They may have different levels of plausibility associated with each of them.

2 *A conditional and subjective quality.* The plausibility of a proposition, meaning an estimation of its truth, is always conditional: it is dependent on every other piece of information we have that we have taken into account. Necessarily then, it is subjective too, since my plausibilities are conditional on my knowledge and yours are conditional on yours. In some fairly simple and therefore rather special circumstance (usually when we are drawing balls from urns) we can agree what the plausibilities are if we can agree what other information (called assumptions) that we will both consider.

3 *Conditional notation.* In order to stress the conditional nature of plausibility we write $A|X$ to represent the plausibility of the proposition that A is true given all background knowledge X, which stands for 'everything that is known *a priori*' when we determine the plausibility that A is true. Now suppose that we gain some new evidence E that was not in X. Then we will write $A|E, X$ to denote the new plausibility that A is true given X *and* given E.

4 *A numerical scale for plausibilities.* Next we wish to measure plausibility on a numerical scale: that is we want to represent the plausibility of the truth of any proposition by a real number on a scale running from zero for false or untrue, up to one for certain or true. Any numerical function measuring plausibility must satisfy some simple constraints on its manipulation and combination though. For example, if there is more than one order in which to assemble hybrid propositions (for example, A and B) then the function and its manipulation must produce consistent results. Under such simple constraints it turns out that there is a unique plausibility function, P, mapping propositions onto the interval zero to one, such that

$$P(A|X) + P(notA|X) = 1, \qquad (A.4)$$

for all propositions A; and

$$P(A \text{ and } B|X) = P(A|B, X) \cdot P(B|X), \qquad (A.5)$$

for all propositions A and B. We call this function P the *probability*.

Notice that we have a sum rule and a product rule. The first condition (A.4) allows us to sum up probabilities over mutually exclusive events, and if, for example a set of m such events, E_i say, are all equally probable (and thus permutable) then each must have probability $P(E_i|X) = 1/m$. The second condition (A.5) says that we must use multiplication to work out probabilities for hybrid propositions (logical 'and's).

5 *Derivation of rules.* The bald fact is that these two rules can be derived from scratch (as solutions to certain functional equations given by Cox and Bretthorst—see Jaynes (2003) [83]), as a consequence of some very reasonable consistency requirements of plausible reasoning, and without any recourse to frequentist repetition of experiments. This is not splitting hairs. Now we are free to apply probability theory to situations which are not equivalent to any repeatable thought experiments, and probability (though necessarily subjective and in context) represents our (un)certainty in any and all propositions that we dare to consider in a self-consistent way.

Have confidence: if we stick to (A.4) and (A.5) and their direct descendants we shall not go wrong.

6 *Combining independent probabilities.* If A is independent of B, then knowledge about B has no impact on our knowledge about A. So $P(A|B, X) = P(A|X)$ and then we see that we can simply multiply up probabilities: $P(A \text{ and } B|X) = P(A|X).P(B|X)$. This is a consequence of the more general product rule (A.5), but in schools it is often taught first!

7 *Bayes' theorem.* This is named after the Reverend Thomas Bayes, 1702–1761. His theory of probability was published posthumously in 1764.

Newly observed evidence, an event E, meaning that 'E is true', updates our estimates of others probabilities in a constant and repeatable way. Suppose we wish to consider the probability that A is true. Then with no further evidence we have $P(A|X)$. Knowing further evidence, E also, we can consider two ways to write $P(A$ and $E|X)$ from (A.5):

$$P(A \text{ and } E|X) = P(A|E, X) \cdot P(E|X) = P(E|A, X) \cdot P(A|X).$$

Hence we get the *posterior* (post E) probability, $P(A|E, X)$ updated from the *prior* probability, $P(A|X)$:

$$P(A|E, X) = P(E|A, X) \cdot P(A|X)/P(E|X). \qquad (A.6)$$

This last equation is known as *Bayes' theorem.* It is often written in different ways. Suppose we have a set of two or more alternative, mutually exclusive, and exhaustive propositions, A_i say. Then for each A_i, $P(A_i|X)$ is the prior probability. Once we know that 'E is true', we can use (A.6) to update these. Since all such equations contain the same denominator, $P(E|X)$ we often just write

$$P(A_i|E, X) \propto P(E|A_i, X) \cdot P(A_i|X). \qquad (A.7)$$

Then if the A_is are exhaustive as well as exclusive, we can always normalize the right-hand sides so that they sum to unity, to obtain an expression that avoids having to deal with the term $P(E|D)$ which we may not know explicitly (though we can know it now!):

$$P(A_i|E, X) = \frac{P(E|A_i, X) \cdot P(A_i|X)}{\left(\sum_j P(E|A_j, X) \cdot P(A_j|X) \right)}. \qquad (A.8)$$

The terms $P(E|A_i, X)$ used in the updating (from prior to posterior for the A_i) can be thought of as *model* terms. Under each separate hypothesis that X and A_i is true, we must have such a model available for calculating this probability for E.

8 *Odds and odds notation.* Sometimes it is easier and mathematically convenient to talk about *odds* rather than probabilities. If p denotes the probability $P(A|E, X)$ for some proposition, A, conditional on X and some event data, E, say, then the odds on A are simply given by

$$O(A|E, \dots, X) \equiv \frac{P(A|E, \dots, X)}{P(\text{not } A|E, \dots, X)} = \frac{p}{(1-p)}.$$

If we know the odds, $O(A|E, \dots, X)$ we can easily find $P(A|E, \dots, X)$ and vice versa.

Just as we have prior and posterior probabilities, we can have the corresponding prior and posterior odds.

Now consider (A.6) as it is written, and again replacing A with its complement *not A*. Then taking the ratio we have Bayes' theorem written for odds:

$$O(A|E, X) = \frac{P(E|A, X)}{P(E|\text{not } A, X)} \cdot O(A|X). \quad (A.9)$$

This formulation is very useful—it avoids the term $P(E|X)$ again since it is a ratio of two equations each with a division by that term. It says the posterior odds are equal to the prior odds multiplied by a ratio of the model terms. If we have a model $P(E|A, X)$ and a single model for $P(E|\text{not } A, X)$ then we can use these directly. If we have split not A up into a greater number of mutually exclusive alternatives then this last formula gets a little more tricky. (See multiple hypothesis testing in Chapter 5).

The model ratio term in (A.9) is often called a *Bayes factor*.

Before we go any further, let us imagine the kind of calculations we might make. If we observe a number of events which are all independent then we may apply (A.9) successively. Each posterior, after each observation, becomes the prior as we update to account for the next observation. Since the events are independent we may simply multiply together all the Bayes factors so as to get the overall posterior odds. Numerically, if the $P(E|A, X)$s are very small then the odds are also small and then this may result in underflows. So it is very good practice to take the logarithm of the odds. See Log Bayes Factors (LBFs) in Section 3.8.

9 *Summaries.* In any problem the priors belong to each of us: they are subjective. We may select them in a number of ways based on what we know. Sometimes we select them to make life easy for ourselves.

However, even when we differ over priors we shall all usually reason in the same direction in the light of new evidence.

The model terms (and the Bayes factors) must be fit for our purpose. This is mathematical modelling: for each hypothesis we must derive or assert a suitable distribution for observable events (over the set of possible values, continuous or discrete) under the assumption that it is true. X, our prior knowledge and skill, is required here.

Finally we have a posterior. A distribution of probabilities over a number of competing hypotheses. It is our job to *summarize* this posterior. This can be done in a number of ways. Perhaps with one or two simple parameters: the modal value (corresponding to the peak, the most likely hypothesis) or the mean/expected value. Or perhaps we should specify a credible set of values/hypotheses. Or even summarize by presenting the whole distribution for inspection.

We are done. Now we are ready to go to work.

Some Examples

Application: Arresting a Suspect

Suppose a man is arrested in a New York neighbourhood suspected of committing a knife murder only one hour earlier. The NYPD believe he has a probability of $p = P(guilty|X)$ of being the murderer, where X represents everything that they know. And the NYPD knows a lot. When he is searched at the station he is found to be carrying a knife. Does that make him more likely to be guilty? At first sight you may think it does, but we really need a little more information to decide the question.

What is P(carrying a knife|not guilty, X)? One in ten men of his age carry a knife like his in the tough neighbourhood where the suspect lives and was picked up. That is our model in this case: P(carrying a knife|not guilty, X)=1/10.

What is P(carrying a knife|guilty, X)? The policemen know that following a knife murder almost all murderers will discard the weapon as soon as they can. After one hour they estimate P(carrying a knife|guilty, X) = 1/50 (so only one in fifty murderers would still have the knife). That is our model in this case. Now we apply (A.8). We have

$$P(\text{guilty}|\text{carrying a knife}, X) = \frac{(1/50)p}{(1/50)p + (1/10)(1-p)} = \frac{p}{5-4p}.$$

Hence for any prior p, the discovery of the knife makes the suspect's guilt less likely.

Of course, we could get a different result if we change the 'models' around a bit. But the lesson is clear. We cannot jump to any conclusion until we estimate the conditional probabilities for the new evidence under the alternative hypotheses, using some distinct 'models' to do so.

Note, if we use (A.9) then things are much simpler in terms of odds:

$$O(\text{guilty}|\text{carrying a knife}, X) = \frac{P(\text{carrying a knife}|\text{guilty}, X)}{P(\text{carrying a knife}|\text{not guilty}, X)}.O(\text{guilty}|X)$$

$$= \frac{1}{5}.O(\text{guilty}|X) = \frac{p}{5(1-p)}.$$

There is a vast literature on Bayesian probability in legal reasoning and other types of inference. The reason for it is precisely because the results can be surprising (apparent paradoxes), and there are lots of cases where inferences are drawn without the full information (necessary to derive the alternative conditional probabilities for the new evidence) being considered (that is, jumping to conclusions).

The Monty Hall Problem

Here is a variant of a rather famous problem. You are on a game show. There are five doors. Behind one door is a prize car; behind the others there is nothing.

The game show host asks you to select two of the doors for yourself (you can keep whatever is behind them). You do so. Then the host, Monty Hall (who knows where the car is hidden), walks up to the remaining three doors and opens up two of them revealing nothing behind them. He then invites you to stick with your original choice of two doors or to swap them both for the single remaining door. You will receive any prize behind the door or the pair of doors that you have after this. Should you swap?

The prior odds that the car is behind one of the two doors you first selected are 2/3 (the prior probability is 2/5 of course). We will use (A.9).

What is the probability that the host could open two of the remaining three doors revealing nothing, assuming you already have already got the car behind one of your doors? It is one: he can easily do it. What is the probability that the host could open two of the remaining three doors revealing nothing, assuming you already have *not* got the car behind one of your doors? It is also one: he can easily do it since he knows which one of the three doors is hiding the car so he can open the other two. Let A denote the proposition that the car is behind either one of the two doors that you initially select. Let E denote the event whereby the game show host opens up the two empty doors.

Then (A.9) become

$$O(A|E, X) = \frac{P(E|A, X)}{P(E|not\, A, X)} . O(A|X) = \frac{1}{1}\frac{2}{3}.$$

So the model ratio term in (A.9) is one: the posterior odds are the same as the prior odds. You still have a 2/5 chance of already having the car. But something has changed. The alternative (*not A*) has now been narrowed down to the car being behind a single door. The probability that the car is behind that door is 3/5 (the only remaining possibility if you have not got the car). Hence you should swap: you should give up your two doors for the final door.

Sloppy reasoning would just say that originally all of the doors had a one in five chance, so why give up two chances for just one? You should stick! Or that at the end we have three doors left; all are equally likely so two chances are better than one. Stick! Yet the single door that is offered in the swap has changed in status. Extra knowledge, that of the host, has been used to select that door for the final threesome, by opening and revealing the other two doors to be empty. That door has been selected in a special way with extra insight, and is not a randomly selected door from the five (as your original two were). Its status has changed. It now has a 3/5 probability of hiding the car because the game show host knew where the car was from the start and he selected it to remain closed.

If the host *did not* know where the car really was then this changes things completely. In that case when he opens up a random pair of doors, selected from the three that you have not selected, he risked revealing the car and you would have lost immediately— Game over!—but fortunately that did not happen. So the probability that the host could open two of the remaining three doors revealing nothing, assuming A, is one. He can easily do it. But the probability that the host could open two of the remaining three doors revealing nothing, assuming *not A*, is 1/3. He had a two thirds chance of revealing the car and ending the game early. So the model ratio term in (A.9) is one divided by 1/3: equal to three. Thus the posterior odds are now two (the prior odds being the same as before, 2/3):

$$O(A|E, X) = \frac{P(E|A, X)}{P(E|not\, A, X)} . O(A|X) = \frac{1}{1/3} . \frac{2}{3} = 2.$$

Hence in the case where the idiotic host, Monty, did not know where the car was you should stick with the pair and not swap.

Example Application: The Swine Flu Test

Suppose there is a new virus such as swine flu that is difficult to detect. It is believed that one in one hundred people have swine flu (SF), but there is also a test (a blood test or something similar). Before anybody takes a test their probability of having SF is therefore 0.01. The test is ninety eight per cent accurate for those with SF: $P(\text{Positive}|SF, X) = 0.98$. The test produces two per cent false positives: $P(\text{Positive}|\text{not } SF, X) = 0.01$.

Suppose he or she takes the test and gets a positive result: what is the probability that he or she has SF now? The prior odds of having SF are $1/99$. Applying (A.9) we have

$$O(A|E, X) = \frac{P(E|A, X)}{P(E|\text{not } A, X)} . O(A|X) = \frac{.98}{.02} \frac{1}{99} \sim \frac{1}{2}.$$

So as a result of the prior population bias, a positive result means that the testee is still twice as likely NOT to have SF than to have it.

This example stresses the importance of *not* simply focussing on the model terms, which make the test look awesome, but to consider the bias on the conditions that each hypothesis is likely to occur.

The transposed Conditional

While writing these notes today, there was a good example of very sloppy plausible thinking in *The Times* (which had to be corrected by the President of the RSS). Drinkers look away now please! *The Times* reported that a high number of patients in high dependency clinics (IHDC) with (near) Liver Failure (LF) were middle class drinkers (MCDs) who drink more than a few glasses of wine at home every evening. The article implied that MCDs had an increased probability of suffering from LF.

But we simply do not have enough information: we know $P(MCD|IHDC, LF, X)$. Suppose nine out ten patients with LF in HDCs are MCDs. We have $P(MCD|IHDC, LF, X) = .9$.

The article sought to imply something about $P(LF|MCD, X)$, transposing the conditional and the consequence; and dropping another condition (IHDC). It implied in particular that $P(LF|MCD, X) > P(LF|X)$, the probability of some random adult suffering with LF. In fact this example is like the knife-crime example discussed earlier, and it is still possible that $P(LF|MCD, X)$ is less than $P(LF|X)$.

Let us make some further assumptions. First we must deal with the IHDC proposition. In odds notation we already have

$$9 = O(MCD|IHDC, LF, X) = \frac{P(IHDC|MCD, LF, X)}{P(IHDC|\text{not}MCD, LF, X)} . O(MCD|LF, X).$$

The middle classes have *sharp elbows*, so our model for tendency is that middle class adults suffering LF are, say, thirty-six times more likely to be IHDC than somebody from the lower classes (notMCD) suffering from LF: the MCDs always visit their doctors when they are unwell. Perhaps the lower classes simply carry on drinking and die, or stay away from HDCs at any rate. We have

$$9 = O(MCD|IHDC, LF, X) = 360(MCD|LF, X),$$

so that

$$O(MCD|LF, X) = 1/4, \quad P(MCD|LF, X) = 1/5.$$

Now directly from Bayes' theorem again:

$$P(MCD|LF, X) = \frac{P(LF|MCD, X)}{P(LF|X)} P(MCD|X).$$

Suppose that $P(MCD|X) = 1/4$, i.e., MCDs make up a quarter of all adults. Then we have

$$\frac{P(LF|MCD, X)}{P(LF|X)} = \frac{P(MCD|LF, X)}{P(MCD|X)} = \frac{1/5}{1/4} = \frac{4}{5}.$$

Here we just made up the extra facts that were missing in the original article. But the point is clear. Even if $P(LF|MCD, X)$ is large we must not confuse it with $P(MCD|LF, X)$. This error is so common that it has a name: the 'transposed conditional'.

Discrete Distributions: Multinomial Models

A multinomial model is simply a way of describing a set of probabilities that some (random) variable is drawn from a discrete set of mutually exclusive and exhaustive alternatives.

Suppose $\{B_j | j = 1, \ldots, m\}$ denotes such a discrete set of alternative possible classes or categories for an observable event, or quantity, b. A multinomial model for the random event b is a set of probabilities $\{P_j | j = 1, \ldots, m\}$, such that

$$P(b \in B_j | X) = P_j, \quad \text{and} \quad \sum_{j=1}^{m} P_j = 1.$$

Suppose that when we observe a number of such events, none of which effect any of the others (so the likelihood of each result is given by our model) then we say they are independent. Suppose that for exactly n_j of these, the result b is in B_j. Then we have

$$P((n_1, n_2, \ldots, n_m) | (P_1, P_2, \ldots, P_m), X) = \prod_{j=1}^{m} P_j^{n_j}.$$

Hence we have a model for the likelihood of observed data, given the set of multinomial probabilities (that must sum to one).

If $m = 2$ then we only talk about $P_1 = p$, say, since $P_2 = (1-p)$, and we have a binomial distribution for the different outcomes (combinations of events) from a number N of experiments.

Continuous Distributions

Often we will wish to run a continuum of hypotheses against one another. We will deal here with a distribution of probable values for some real parameter λ: the generalization to higher dimensions or other types of state space is obvious in almost all cases.

Suppose we have some real constant λ that is unknown. Let S be any subset of the real line and define the corresponding hypothesis, H_S, via

$$H_S = \text{``}\lambda \in S\text{''}.$$

Then we introduce a *probability density function*, often called a *pdf*, $f(\lambda|X)$. This is a non-negative real-valued function, and

$$P(H_S|X) = \int_S f(\lambda|X)d\lambda.$$

Each hypothesis is intimately linked to its set of hypothesized values, S. Clearly hypotheses are mutually exclusive if the intersection between their sets is empty (or of measure zero).

If S contains all possible values (the entire support of f) then $P(H_S|X) = 1$. Hence f must be of unit mass as well as non-negative.

Bayes' theorem still applies, as it must. So for any new event or evidence, E, we can write

$$f(\lambda|E,X) = \frac{P(E|\lambda, X)f(\lambda|X)}{P(E|X)}.$$

This last follows by applying Bayes' theorem to $P(H_S|X)$, and its posterior counterpart, $P(H_S|E,X)$, and observing that S can be chosen arbitrarily.

Normalizing probability density functions is tedious.

We know that the total mass of any posterior pdf must be unity. So often we will prefer to deal with *non-normalized density functions*, and write simply

$$f(\lambda|E,X) = P(E|\lambda, X)f(\lambda|X),$$

with the running proviso that we will own up and normalize such fs whenever that is required.

Of course we also may stray into the area of fs which are of infinite mass (not integrable) if we wish. These are called *improper density functions*. For example $f(\lambda|X) = 1$ for all real λ, or $f(\lambda|X) = 1/\lambda$ for all positive λ. This idea can indeed be very useful, since a posterior pdf may be integrable (thus proper) even when the prior is improper.

Suppose we have absolutely no prior information to constrain our thoughts about λ: how will we set our prior then? Improperly? This is a subjective issue, a constant prior pdf for a variable λ will be constant, yet it induces a prior pdf for, say, $\log \lambda$ that is certainly not constant. Which variable are you relatively indifferent about? On which scale will you speed out your prior mass uniformly? Let us put this issue into X for now. Interested readers should consult Jaynes and Bretthorst (2003) [83], especially on the Jeffreys prior.

As successive events are observed we will evolve a posterior pdf. So a prior pdf should not rule any values out that we might have otherwise accepted later. Eventually, with enough observations, the prior becomes less important. Again consult the references, especially Jaynes and Bretthorst (2003) [83], on this aspect.

Now consider an unknown real parameter λ which is used to model some observable z. We will have a model $P(z|\lambda, X)$ which yields

- a probability density function for z, given a value for λ, if z is continuous
- a multinomial for z, given a value for λ, if z is categorical or discrete.

For example, our model may assert that z is normally distributed about λ with unit variance, say. So given λ we can estimate a possible values for z.

Now suppose we observe an actual value z^*. What does this tell us about λ? Well the event E is simply the observation itself: that z lies with any set S containing z^*.

So we have

$$f(\lambda|E, X) \propto P(E|\lambda, X)f(\lambda|X) = \int_S P(z|\lambda, X)dz\, f(\lambda|X).$$

But S was arbitrary, about z^*. So we have

$$f(\lambda|E, X) \propto P(z^*|\lambda, X).f(\lambda|X).$$

Again we can normalize if we wish and write

$$f(\lambda|E, X) \propto \frac{P(z^*|\lambda, X)f(\lambda|X)}{\int P(z^*|\lambda, X)f(\lambda|X)d\lambda}.$$

Now let us start with some pleasant distribution for our prior $(f(y|X))$, and some well-chosen model $P(z|\lambda, X)$ for our observable, and suppose that we observe a set of m independent measurements: $E = \{z_1, \ldots, z_m\}$.

Then we will have the (improper) posterior

$$f(\lambda|E, X) = P(z_1|\lambda, X.P(z_2|\lambda, X) \ldots P(z_m|\lambda, X)f(\lambda|X).$$

This will soon get rather hairy! When the models are algebraically complicated these become difficulty to deal with. How will we summarize the posterior? We will probably struggle to find even its mean or mode.

There are two ways around this: a traditional approach as outlined in Appendix 0.20, which uses some trickery to reduce the amount of algebra, by a pragmatic choice of the prior; and the use of the computer to summarize and sample from the posterior, as written. This last possibility is only thirty or forty years' old and is the subject of much progress, typically referred to as Monte Carlo Markov Chain methods.

Algebraic Convenience: Conjugate Priors

In the days before computers could be used to summarize and sample from posteriors, in order to avoid excessively complicated functions a rather useful practice was developed: the use of *conjugate priors*.

The central and elegant idea is that if the model $P(z|\lambda, X)$ is given, then rather than choose any prior, $f(\lambda, X)$ for λ, if we had no better reason then we could make life very easy for ourselves by choosing f from a particular family of functions, so that the posterior would be from the same family. Such a family of functions is called a *conjugate prior* distribution for the chosen model distribution.

Let $F(\lambda|\theta)$ be a family of pdfs (normalized or not) for λ parameterized by θ. Then F is conjugate to the model, $P(z|\lambda, X)$, if the posterior is given by

$$F(\lambda|\hat{\theta}) = P(z|\lambda, X).F(\lambda|\theta),$$

where $\hat{\theta} = \hat{\theta}(\theta, z)$, is some well-defined function. Notice that if the model is given, then this last is a functional equation for F and $\hat{\theta}$.

For example, suppose z is a multinomial variable, with m categories. Then let λ be the vector $(P_1, P_2, \ldots, P_m)$ of the unknown probabilities in our multinomial model. Then λ lives on the simplex: $\lambda \geq 0$, and $\lambda^T \mathbf{s} = 1$, where $\mathbf{s} = (1, 1, \ldots, 1)^T$.

As we observe instances of the multinomial variable z we will change our opinion as to where the $\lambda = (P_1, P_2, \ldots, P_m)$ may lie.

Now for any real $\mathbf{w} = (w_1, w_2, \ldots, w_m)^T \geq 0$ let

$$G(P_1, P_2, \ldots, P_m, \mathbf{w}) = \prod_{i=1}^{m} P_i^{w_i},$$

be defined on the simplex $\{P_i \geq 0, \sum P_i = 1\}$. Strictly speaking we should have normalized G so that it integrates to one, but we can proceed with this improper form. Note that on this simplex $G(P_1, P_2, \ldots, P_m, \mathbf{w})$ has a maximum (modal value) at $\mathbf{w}/\|\mathbf{w}\|$ (hint: use Lagrange multiplier to maximize G while constraining to the simplex). If $\mathbf{w} = 0$ then G is uniform.

Then suppose our prior 'insight', X, allows us to select some nonnegative real values for $\mathbf{w}$ and to take the prior

$$f((P_1, P_2, \ldots, P_m), X) = G(P_1, P_2, \ldots, P_m, \mathbf{w}).$$

Now suppose that we observe some evidence, E, containing a, sets, of independent instances of the categorical variable z with exactly n_i of them within C_i.

Let $\mathbf{n} = (n_1, \ldots, n_m)^T$, then we have the posterior

$$f((P_1, P_2, \ldots, P_m)|E, X) = \prod_{i=1}^{m} P_i^{n_i}.G(P_1, P_2, \ldots, P_m, \mathbf{s}) = G(P_1, P_2, \ldots, P_m, \mathbf{n} + \mathbf{w}),$$

which is of the same family as the prior. Hence G is the conjugate prior for the multinomial.

For many, many observations the posterior becomes peaked around its modal value at $\sim \mathbf{n}/\|\mathbf{n}\|$.

Note that if $m = 2$ and the multinomial is a binomial we usually write $P_1 = p$ and $P_2 = 1 - p$, and abuse the notation to write

$$G = G(p, \mathbf{w}) = p^{w_1}(1 - p)^{w_2}.$$

In this case, G is called a beta distribution (when normalized) and will be discussed in more detail in Appendix 0.21.

If *a priori* we think that a coin is likely to be a fair one we might select $w_1 = w_2 = 10$. But after observing Q consecutive heads (C_1), and no tails, then we have the posterior $G(p, (Q + 10, 10)) = p^{Q+10}(1 - p)^{10}$ which has a modal value at $p = (Q + 10)/(Q + 20)$.

In our next example of a conjugate prior suppose that we will observe some real quantity z. As a model for z we will choose a Poisson distribution with intensity λ, some unknown parameter. We have the model distribution

$$P(z|\lambda, X) = \lambda e^{-\lambda z}.$$

Suppose next that we feel able to choose the prior

$$f(\lambda|x) = \lambda^a e^{-\lambda b},$$

for some non-negative constants a and b. Again we have not bothered to normalize this f. It has a modal peak at $\lambda = a/b$ (Calculus!).

Then observing a single value of z, say at z^*, we have the posterior

$$f(\lambda|z^*, x) = P(z^*|\lambda, X).f(\lambda|X) = \lambda^{a+1} e^{-\lambda(b+z^*)}.$$

If we continue adding further independent observations, so that there are m in total, then the modal value approaches the inverse of the mean of the observed z-values. We will have $\lambda_{mode} = (1 + a/m)/(\bar{z} + b/m)$.

The two-parameter family $\lambda^a e^{-\lambda b}$ is the conjugate prior for the Poisson distribution.

In Section 6.4 we show that the multivariate Gaussian distribution is conjugate to itself.

Laplace and Laplace's Law of Succession

We often want to estimate probabilities by counting instances of different types of event. We should guard against setting event probabilities to one or zero, even if we have (so far) always observed them or have never yet observed them. The fact that we subjectively think alternative event are possible means we have to allow at least some small amount of probability.

Suppose we have an urn containing some red and some white balls. All probability theorists just love urns containing balls—it is the fault of the Bernoulli's (look them up). Our prior information about them is that the balls are always well mixed within such urns and each has the same probability of being sampled though a single 'draw'. But in this case we do not know how many balls of each type are in the urn, though we will assume that both types of balls are contained within the urn. We draw a ball at random from the urn and examine its colour. We replace it and draw again. Suppose after N such draws, we have drawn a red ball exactly n times and a white ball exactly $m = N - n$ times. What is the probability p that the next ball to be drawn (and any ball subsequent ball drawn independently) will be red?

We do not know p at this stage. But we can express our knowledge about the possible values that the parameter p might take by using a *probability density distribution*, $f(p|x)$ say. As before, this is a non-negative function defined over all of the possible values of p (in this case (0,1)) such that the probability that the true value of p lies within the interval between two constants $a > 0$ and $b > a$ (but less than one) is given by the integral of the density distribution over the interval:

$$P(a < p < b|X) = \int_a^b f(p|X)dp.$$

Note that we are using a probability distribution for the value of the parameter, p, which is itself a probability.

When probability distributions for some parameter are integrated over all of its possible values, we must obtain unity. In our present case we have always

$$P(0 < p < 1|X) = \int_0^1 f(p|X)dp = 1.$$

If f has a maximum at some value p^*, then this represents the maximum likelihood value, or mode value, for p. If the distribution is narrow and highly spiked then we must think that the true value for p lies very close to the mode value. Conversely, if the distribution is rather flat with a large variance we must be unsure as to where the true value lies, and have no great preference for one estimate over another.

For any function of p, say $G(p)$ which is given, the *expected value* for G is given by:

$$\langle G|X \rangle = \int_a^b G(p)f(p|X)dp.$$

In particular, our expected value of p itself is just:

$$\langle p|X \rangle = \int_a^b pf(p|X)dp,$$

and our expected value of any power p^q is just:

$$\langle p^q|X \rangle = \int_a^b p^q f(p|X)dp.$$

Hence the variance of the distribution is

$$\sigma^2 = \langle p^2|X \rangle - \langle p|X \rangle^2 = \int_a^b (p - \langle p|X \rangle)^2 f(p|X)dp.$$

Now let us return to thinking about our urn filled with some red and some white balls, where p is the probability that any ball to be drawn will be red. As our information about the possible true value of p changes, this will change the distribution, and hence our estimates. This information changes each time we draw a single ball.

At the start of this thought experiment we know only that p lies between zero and one. Let us assume a prior distribution, $f_0(p|X) \equiv 1$, that is uniform (and equal to one) for all values between zero and one. That is, before any balls have been drawn, we will assume that the probability that the model parameter p lies between $a > 0$ and $b > a$, but less than one, is

$$P(a < p < b|X) = \int_a^b f_0(p|X)dp = (b - a).$$

Using Bayes' formula after drawing n reds and m whites we have a posterior distribution

$$f(p|(n, m), X) \propto P((n, m)|p, X)f_0(p|X).$$

But $f_0 \equiv 1$, so by integrating both sides of this equation we have

$$f(p|(n, m), X) = \frac{P((n, m)|p, X)}{\int_0^1 P((n, m)|p, X)dp}.$$

Now if p is considered known, then the probability of drawing each red independently is p and the probability of drawing each white independently is $1 - p$. Therefore we have the simple 'model' for the event (n, m):

$$P((n, m)|p, X) = p^n(1 - p)^m.$$

This is called the 'binomial distribution': the probability of drawing the various results (n, m), given a value for p. Considered as a function of p, for (n, m) given, it is called the 'beta distribution'. Above we see that our posterior distribution for p is just a normalized form of this function:

$$f(p|(n, m), X) = \frac{p^n(1 - p)^m)}{\int_0^1 p^n(1 - p)^m dp}. \tag{A.10}$$

So what is our expected value $\langle p|X \rangle$ for p (the expected probability of drawing a red ball in the next draw)? We have the ratio of two integrals

$$\langle p|X \rangle < p|X >= \frac{\int_0^1 p^{(n+1)}(1 - p)^m dp}{\int_0^1 p^n(1 - p)^m dp}. \tag{A.11}$$

Using integration by parts again and again we have:

$$\int_0^1 p^s(1-p)^r)dp = \frac{r}{(s+1)} \int_0^1 p^{(s+1)}(1-p)^{(r-1)}dp$$

$$= \frac{r(r-1)}{(s+1)(s+2)} \int_0^1 p^{(s+2)}(1-p)^{(r-2)}dp$$

$$\vdots \qquad\qquad\qquad\qquad \text{(A.12)}$$

$$= \frac{r(r-1)\dots 2.1}{(s+1)(s+2)\dots(s+r-1)(s+r)} \int_0^1 p^{(s+r)}dp$$

$$= \frac{r!s!}{(s+r+1)!}.$$

Using (A.12) twice in (A.11), we obtain:

$$\langle p|X\rangle = \frac{n+1}{n+m+2}. \qquad\qquad \text{(A.13)}$$

This (and its generalizations) is called *Laplace's Law of Succession*. Laplace first gave it in 1774 and it has played a major role in the story of Bayesian inference. Many books on probability theory try to ignore it, but it is remarkably useful for our purposes.

We can also see how good an estimate for the true value the expectation $\langle p|X\rangle$ might be by considering the standard deviation, σ, the square root of the variance. We have, again using (A.12), and some rearrangement (exercise):

$$\sigma^2 = \frac{\langle p|X\rangle(1-\langle p|X\rangle)}{(n+m+3)}.$$

As is often the case, the standard deviation goes to zero like $(n+m)^{-1/2}$: increase the sample size by one hundred to decrease the error by ten. But we are content with (A.13) even when $n+m$ is small, since it represents our expectation given our prior knowledge and the data we have, and now we also have an estimate, σ, as to how wide the resulting posterior distribution is.

Now let us consider some of the surprising things about Laplace's Law. First, it does not depend on the number of balls in the urn, which we do not know. In fact the urn is just a mechanism for producing results: we simply generate a red ball with some unknown probability p. What can we say about p? The details of the shape of the urn or the number of balls in it do not matter. In our applications we will frequently need to estimate parameters based on a limited number of calibration results. Then we can use Laplace's Law.

Second, it is never equal to zero or one. Even if we have never seen a white ball and have drawn N red balls in succession, $\langle p|X\rangle = (N+1)/(N+2)$. There is always a chance (based on our prior information that white balls can be in such urns) that the next draw will be white.

If we have two identical urns, but with distinct mixes of balls, and have drawn one hundred red balls (no whites) from urn one and 1,000 red balls (no whites) from urn two, we can say that if we are to see a white ball appearing it is much more likely to come from urn one than urn two, because for the latter we have carried out more draws without a white appearing. This is consistent with our common sense.

As an exercise, find the modal value for p. This is where $p^n(1 - p)^m$ has a maximum. Consider how this differs from the expected value given by (A.13). What if we have made N draws and all of them are red balls?

The *general case of Laplace's Law of Succession* is given as follows. (Just think of urns containing balls of many, K, distinct colours.)

Consider a mechanism where there are a number, K, of possible mutually exclusive types of result, A_k say, for $k = 1, \ldots, K$; each generated by a corresponding causal process that remains constant. We suppose that each type of result, A_k, occurs with a probability p_k where $\sum_{k=1}^{K} p_k = 1$. Then suppose we have obtained N results, or have made N observations, and obtained A_k exactly n_k times (of course $\sum_{k=1}^{K} n_k = N$). Then what estimates can we have for the p_k?

This involves making some tricky integrals over sets in K-dimensional space. But the argument is analogous to that given above for the case $K = 2$. We obtain

$$\langle p_i | X \rangle = \frac{n_i + 1}{N + K}. \tag{A.14}$$

This generalizes (A.13) (simply put $K = 2, p_1 = p, n_1 = n, N = n + m$).

Equation (A.14) is very useful to us, because whenever we want to express a probabilistic model (for use in Bayes' theorem) as a multinomial model, we need to estimate the probabilities based on some calibration data sets. These estimates avoid absolute zero (impossibility) and absolute unity (certainty), and allow for possibilities barely, or indeed never yet, observed in the data.

Now we come to another very interesting feature. (A.14) depends upon K, the number of results that are thought to be possible, even if some have never yet been observed. For example, suppose we believe that our urn contains yellow, red, and white balls, and we take ten draws obtaining four white, six red, and zero yellow. Then we have

$$p_{red} = 7/13, \ p_{white} = 5/13, \ p_{yellow} = 1/13.$$

Next, suppose that we are told that sometimes the urns may contain green balls also (so X is changed). If we believe this then we must recompute with $K = 4$. So

$$p_{red} = 7/14, \ p_{white} = 5/14, \ p_{yellow} = 1/14 \text{ (and } p_{green} = 1/14).$$

Hence p_{red} depends not just on (n_{red} and N) but on the number of possibilities we are prepared to entertain.

Suppose next we are convinced that we can observe and discriminate between 1,000 different colours and hues.

But perhaps this is a desirable feature since if we want to admit the possibility of rare types of events (for which there is little or no hard evidence), we have to allocate a small chance to their occurrence. Before any data are observed, if we have no prior bias then all such events are to be deemed equally likely. Hence Laplace moves us seamlessly from the prior state of knowledge (all equally possible), through the small data situation, and on to the large data situation.

Unfortunately Laplace did not really help himself popularize this idea by applying his law to calculate the probability that the sun will rise tomorrow given that it has done so for 5,000 years (without assuming any prior knowledge of the workings of the solar system). The odds on tomorrow's sunrise are

$$p/1 - p = 1.83M.$$

Good. But any additional information will also alter our knowledge, and Laplace himself knew a great deal about the mechanics of the heavens, so it is hard for us to put that aside. Our knowledge of what may or may not cause the failure of the sun to rise means that this is not a good example, and surely Laplace only meant this to represent a statement of knowledge given only the fact of n successes in N (independent) trials.

The debate that this formula stirred up has led to over 200 years of objections and woolly thinking. Nevertheless, we stress that this formula is an extremely useful rule for us, especially since it converts data into estimates for probabilistic model parameters that we can employ within models for random (biased) processes. In particular, we can use it to calibrate multinomial models and, by extension, Markov models.

SOLUTIONS TO EXERCISES

Chapter 1

1.1 Suppose not. Let $(\lambda, \mathbf{v})$ be a complex eigen-pair. Then both $\lambda \mathbf{v} = A\mathbf{v}$ and $\bar{\lambda}\bar{\mathbf{v}} = \bar{A}\bar{\mathbf{v}}$. Multiplying the first of these by $\bar{\mathbf{v}}^T$ we obtain

$$\lambda \bar{\mathbf{v}}^T \mathbf{v} = \bar{\mathbf{v}}^T A \mathbf{v} = (\bar{\mathbf{v}}^T \bar{A}^T)\mathbf{v} = (\bar{A}\bar{\mathbf{v}})^T \mathbf{v} = \bar{\lambda}\bar{\mathbf{v}}^T \mathbf{v}.$$

So $\bar{\lambda} = \lambda$.

1.2 $(I - \alpha A)^{-1}$ exists since $1/\alpha$ is greater than the spectra radius. Now we may write $(I - \alpha A)^{-1} = S$ where

$$S = I + \alpha A + \alpha^2 A^2 + \alpha^3 A^3 + \cdots$$

This follows since the sum of the geometric series on the right exists, and we can check directly that it satisfies $(I - \alpha A)S = I$. For each i, j, $S_{i,j}$ is a sum of non-negative terms. If it is strictly positive then there must be a lowest power, $k(i, j)$ of A such that $A^{k(i,j)} > 0$. If this is true for all (i, j), then A is irreducible. Conversely if A is irreducible, then for all (i, j) there must be terms in $S_{i,j}$ that are non-zero.

The matrix exponential is defined by the power series

$$\exp(A) = I + A + \frac{A^2}{2!} + \frac{A^3}{3!} + \frac{A^4}{4!} + \cdots.$$

The same argument applies. A is irreducible if, and only if, the matrix $\exp(A)$ is strictly positive.

1.3

$$A^2 = \begin{pmatrix} B_1 B_2 & 0 \\ 0 & B_2 B_1 \end{pmatrix}.$$

So if $A_{i,j} = 0$ then $(A^2)i, j \neq 0$ Hence A is irreducible.

1.4

(a) From the definition:

$$Q(A) = U(q_0 I + q_1 \Lambda + \cdots + q_m \Lambda^m) U^T.$$

But $A^r = U \Lambda^r U^T$ for $r = 1, \ldots, m$.

(b) Apply (a) to A^{-1} and then use $A^{-r} = U \Lambda^{-r} U^T$ for $r = 1, \ldots, m$.

(c) It is clear that $U Q(\Lambda)^{-1} U^T$ is the inverse of $Q(A) = U Q(\Lambda) U^T$, by direct multiplication.

1.5 It is clear that if S exists then $(I - \alpha A) S = I$: so the result follows. On the other hand, directly we have

$$S = U(I + \alpha \Lambda + \alpha^2 \Lambda^2 + \alpha^3 \Lambda^3 + \cdots) U^T = U(I - \alpha \Lambda)^{-1} U = U f(\Lambda) U,$$

where we take $f(x) = (1 - \alpha x)^{-1}$, which is clearly well defined since all of the eigenvalues are inside the disc $x = 1/\alpha$.

1.6 This network is not edge independent. If we think of the $k > 2$th joiner then each of the $k - 1$ existing employers has a $1/(k - 1)$ chance of being the senior buddy. But if we find out that in fact employee number 1 is the buddy, then immediately there can be no attachment between k and the other $k - 2$ employees.

When the kth person arrives, they have one buddy and all subsequent employers $j = k + 1, \ldots, K$ become k's buddy with probability $1/(j - 1)$. Let d_k denote the degree of the kth employee. Thus

$$\langle d_k | K \text{ employees}, X \rangle = 1 + \sum_{j=k+1}^{K} 1/(j - 1).$$

As K tends to ∞ this becomes unbounded: for each k fixed it goes off like the harmonic series

$$1 + 1/2 + 1/3 + 1/4 \cdots.$$

The harmonic series may be counterintuitive to non-mathematicians first encountering it, because it is a divergent series though the limit of the individual term is zero. In this case, all staff will approach an infinite number of buddies even as the number of staff itself approaches infinity.

1.7 Clearly none of these situations is edge independent. Once you know the identity of a trouper's corporal, a swan's mate, a dancers' partner (even for just one dance) it precludes or reduces the possibilities of other alternatives (which were alive *a priori*).

Non-independence becomes less of an issue as the number of dances increases, provided we are only told about a fixed few of the dancer's partners, although the evening becomes more exhausting.

1.8 For a clique, degree distribution is one at $n-1$ and zero otherwise.

For a $CL(n,k)$, all vertices have degree $2k$ $(<n)$ so the degree distribution is one at $2k$ and zero otherwise.

On a lattice $L(n,k)$ there are $n-2k$ vertices with degree $2k$,

2 vertices with degree $2k-1$,

2 vertices with degree $2k-2$,

...2 vertices with degree k.

Thus the degree distribution $P(j)$ is $2/n$ at $j=k,\ldots,2k-1$ and $(n-2k)/n$ at $j=2k$: zero otherwise.

1.9 A vertex with degree k in G has degree $n-1$ in the clique (there are $n-1$ possible edges) and thus $n-1-k$ in G'. So if $P'(k')$ denotes the degree distribution for G', we have $P'(k') = P(n-1-k')$.

1.10 The prior estimate (before we are given the information that v is connected to another vertex v_0) is that v has degree k with probability $P(k)$. Let E denote the proposition that v is connected to v_0 and H_k denote the proposition that the degree of v is equal to k. Then Bayes theorem says that the posterior satisfies:

$$\tilde{P}(k) = P(H_k|E,X) \propto P(E|H_k,X)P(H_k|X) = P(E|H_k,X)P(k).$$

But $P(E|H_k,X) = k/(n-1)$ is the best information we have, since we have no information to distinguish v_0. Thus

$$\tilde{P}(k) \propto kP(k).$$

The normalization means that $\sum_k \tilde{P}(k) = 1$ so

$$\tilde{P}(k) = \frac{k}{z}P(k),$$

where $z = \sum_k kP(k)$ is the expected degree for vertices in G.

1.11 The mean, or expected value, $\langle k|X\rangle$, is given by

$$\sum_{k=0}^{\infty} \frac{k\lambda^k e^{-\lambda}}{k!} = \sum_{k=1}^{\infty} \frac{k\lambda^k e^{-\lambda}}{k!}$$

$$= \sum_{k=1}^{\infty} \frac{\lambda^k e^{-\lambda}}{(k-1)!} = \lambda \sum_{k=1}^{\infty} \frac{\lambda^{(k-1)} e^{-\lambda}}{(k-1)!} = \lambda.$$

The variance, $\langle (k-\langle k|X\rangle)^2|X\rangle$, is given by $\langle k^2|X\rangle - \langle k|X\rangle^2$. Here $\langle k|X\rangle = \lambda$. Now

$$\langle k(k-1)|X\rangle = \langle k^2|X\rangle - \lambda.$$

But the left-hand side is

$$\sum_{k=0}^{\infty} \frac{k(k-1)\lambda^k e^{-\lambda}}{k!} = \sum_{k=2}^{\infty} \frac{\lambda^k e^{-\lambda}}{(k-2)!} = \lambda^2.$$

Hence

$$\lambda^2 = \langle k^2 | X \rangle - \lambda.$$

But the variance is given by $\langle (k-\lambda)^2 | X \rangle$ which is thus equal to

$$\langle k^2 | X \rangle - \langle k | X \rangle^2 = \lambda$$

as required.

1.12 A circular lattice and its complement: $CL_{12,3}$ and $1\text{-}CL_{12,3}$.

1.13 Take the logarithm of $\mathcal{G}$:

$$\log \mathcal{G}(x) = 2 \sum_{j=1}^{\infty} \log(xf(j) + (1 - f(j))).$$

Now differentiate

$$\frac{\mathcal{G}_x(x)}{\mathcal{G}(x)} = 2 \sum_{j=1}^{\infty} \frac{f(j)}{xf(j) + (1 - f(j))}.$$

Set $x \to 1$.

1.14 For a generating function

$$\mathcal{G}(x) = \sum_{k=0}^{\infty} P_k x^k,$$

we have the expected degree given by $\mathcal{G}'(1)$, and the kth derivative of $\mathcal{G}$ at zero divide by $k!$ is P_k the probability that a vertex has degree k.

$$G'(x) = \frac{\sin\left(\sqrt{b-bx}\right)\left(\sqrt{b-bx}\sin\left(\sqrt{b-bx}\right) + b(x-1)\cos\left(\sqrt{b-bx}\right)\right)}{b(x-1)^2 \sqrt{b-bx}}.$$

So $\mathcal{G}'(1) = b/3$.

Now $P_0 = G(0) = \dfrac{\sin^2\left(\sqrt{b}\right)}{b}$. $P_1 = \dfrac{\sin\left(\sqrt{b}\right)\left(\sqrt{b}\sin\left(\sqrt{b}\right) - b\cos\left(\sqrt{b}\right)\right)}{b^{3/2}}.$

For the very, very, brave (or Maple/Mathematica users) we have

$$P_2 = -\frac{-4\sqrt{b} + 5b\sin\left(2\sqrt{b}\right) + 2(2-b)\sqrt{b}\cos\left(2\sqrt{b}\right)}{8b^{3/2}},$$

and

$$P_3 = \frac{24\sqrt{b} - (33 - 4b)b\sin\left(2\sqrt{b}\right) - 6(4 - 3b)\sqrt{b}\cos\left(2\sqrt{b}\right)}{48b^{3/2}}.$$

1.15 This is straightforward if you imagine the adjacency matrix in the subgraph centred at v_i as described. There are $2k(2k+1)/2$ elements in that upper triangle of which $2k$ involve an edge to or from the central vertex v_i. Thus there are $2k(2k+1)/2 - 2k = 2k(2k-1)/2$ pairs of adjacent vertices. But there is a furthest triangle, at the top right, containing $k(k-1)/2$ zeros. So there remain $(4k^2 - 2k - k^2 + k)/2$ ones in the upper triangle, equal to $(3k^2 - k)/2$ as required. Hence $C = (3k^2 - k)/(4k - 2)$ in this case.

1.16 First part is just counting. The ikth off diagonal term is the product of the ith row/column with the jth row/column. Thus it counts the number of common neigbours v_j. Hence N_2 is the sum of such terms and thus all possible connected triples, starting at both ends.

The diagonal term $(A^3)_{ii}$ counts the number of walks of length three from v_i to v_i including the reverse directions. They are all distinct. So N_3 counts all triangles, including the directions and starting at all three vertices.
$C = N_3/N_2$ is obvious.

Chapter 2

2.1 First we have

$$\mathbf{w} = \mathbf{r} - \mathbf{s} = ((I - \alpha A)^{-1} - I)\mathbf{s} = \alpha A(I - \alpha A)^{-1}\mathbf{s} = \alpha A\mathbf{r}.$$

Let the eigen–pairs for A be denoted by $(\lambda_j, \mathbf{v}_j)$ where $j = 1, \ldots, n$, and $j = n$ corresponds to the Perron–Frobenius pair, where $\lambda_n = \lambda_{PF} > 0$ is the spectral radius of A. Introduce the expansions $\mathbf{w} = \sum_{j=1}^{n} w_j \mathbf{v}_j$ and $\mathbf{s} = \sum_{j=1}^{n} s_j \mathbf{v}_j$. Then equating coefficients we have $w_j = \alpha \lambda_j(w_j + s_j)$, so that $w_j = \frac{\alpha \lambda_j}{1 - \alpha \lambda_j} s_j$. Hence as $1/\alpha \to \lambda_n$ for above $w_j \to \infty$. Thus $\mathbf{w}$ become asymptotic to the $\mathbf{v}_n$ direction, and hence the relative ordering to the vertices via Katz centrality become equivalent to that achieved via eigenvector centrality.

2.2 We write $A = U\Lambda U^T$, where U is a unitary matrix with columns given by eigenvectors of A, and we have the real diagonal matrix of eigenvalues: $\Lambda = \text{diag}(\lambda_1, \lambda_2, \ldots, \lambda_n)$, (and $\lambda_n > 0$ is the Perron–Frobenius eigenvalue). Then we can write $B = U\Lambda^{1/2}U^T$ where $\Lambda^{1/2} = \text{diag}(\lambda_1^{1/2}, \lambda_2^{1/2}, \ldots, \lambda_n^{1/2})$. Since there are two roots for each eigenvalue, there are 2^n possible matrix square roots, B.

2.3 This is just the remainder theorem for polynomials. Let $Q(\lambda)$ denote the characteristic polynomial of A of degree n. Then for all $P(\lambda)$ there is a polynomial of degree less than or equal to $n-1$, $R(\lambda)$ say, and a polynomial $S(\lambda)$ such that $P(\lambda) = Q(\lambda)S(\lambda) + R(\lambda)$. At the eigenvalues $Q(\lambda_k) = 0$ thus $P(\lambda_k) = R(\lambda_k)$. Now

$$P(A) = UP(\lambda)U^T = UR(\Lambda)U^T.$$

Thus once A is diagonalized we can construct any polynomial in A by first reducing it to one of degree at most $n-1$, and then evaluating it at the eigenvalues of A.

Notice that any two functions that happen to agree on the spectrum of A produce identical images of A, as matrix functions.

2.4 First part is tedious calculation. We obtain

$$
\mathcal{Q} = \begin{pmatrix} 1.57351 & 0.987937 & 0.765986 \\ 0.867483 & 1.59238 & 0.746939 \\ 1.01152 & 1.1073 & 1.54014 \end{pmatrix}.
$$

Clearly the largest row sum is that from row three. The largest column sum is from column two.

2.5 We have the expansion

$$
\mathcal{Q} = I + \alpha \sum_{k=1}^{K} A_k + \alpha^2 \sum_{k=1}^{K} \sum_{k'=k}^{K} A_k A_{k'} + O(\alpha^3).
$$

Thus since all of the As are symmetric:

$$
\mathcal{S}(\mathcal{Q}) = I + \alpha \sum_{k=1}^{K} A_k + O\alpha^2),
$$

and

$$
2\mathcal{AS}(\mathcal{Q}) = \alpha^2 \sum_{k=1}^{K-1} \sum_{k'=k+1}^{K} (A_k A_{k'} - A_{k'} A_k) + O(\alpha^3).
$$

Thus any anti-symmetric terms only arise through the interaction of matrices from at least two distinct time steps (k and $k' > k$). Moreover they will be $O(\alpha^2)$, whereas the total *size* of $\mathcal{S}(\mathcal{Q})$ (and hence of $\mathcal{Q}$) is $1 + O(\alpha)$.

Chapter 3

3.1 In that case we have the probability that the walk remains at vertex i in time δt is $1 - \kappa \delta t$, and the probability of moving from i to any specific connected j is $\kappa \delta t / d_i$. Putting all of this together, for all vertices, and using the fact that the walk's position at time t is distributed, we get

$$
\mathbf{z}(t + \delta t) = \kappa \delta t A D^{-1} \mathbf{z}(t) + (I - \kappa)\mathbf{z}(t).
$$

So

$$
\frac{\mathbf{z}(t)}{dt} = -\kappa(I - AD^{-1})\mathbf{z}(t).
$$

3.2 Here $\langle A_{k+1}|X \rangle$ is obtained by using the substitution $\langle A_k|X \rangle = p_k \mathbf{1}$ in (3.3) in place of A_k. Notice that trivially $\mathbf{1} \circ \mathbf{1} = \mathbf{1}$. Of course, multiplying probabilities together and thus using $\langle A_k^2|X \rangle = \langle A_k|X \rangle^2$ is the key trick. Yet (for the off-diagonal terms) this only combines probabilities for the existence of separate edges, which we have assumed to be independent, so their probabilities can be multiplied. Thus we get p_k^2 for each possible two-edge walk between vertex v_i and vertex v_j there are $n-2$ such walks, independent of each other (so we may sum their probabilities). Hence $\langle A_k^2 | X \rangle = (n-2)r_k^2 \mathbf{1}$ as required.

3.3 Let $p^* = 1/2 + \sqrt{1/4 - \gamma/\epsilon(n-2)}$. At the upper stable equilibrium, the expected number of friendships in $A_k \approx p^* \mathbf{1}$ is given approximately by

$$N_f = \frac{n(n-1)}{2} p^*.$$

What is the probability that $A_{k+1} = 0$, given $A_k \approx p^* \mathbf{1}$?

$$P(A_{k+1} = 0 | A_k = p^* \mathbf{1}, X) = \gamma^{N_f} (1 - (n-2)\epsilon p^{*2})^{\frac{n(n-1)}{2} - N_f}.$$

If $n = 50$, $\epsilon = 1/48$, and $\gamma = 0.2$ then there are 1,225 possible friendships. But $p^* = 0.5 + \sqrt{0.05} = 0.723607$ so $N_f = 886.418$. Thus

$$P(A_{k+1} = 0 | A_k = p^* \mathbf{1}, X) = 2.43 \times 10^{-729}.$$

It will happen eventually.

3.4

(a) We must have $Q\mathcal{F}(A)Q^T = \mathcal{F}(QAQ^T)$, so the result follows by premultiplying by Q^T and post multiplying by Q and using $Q^T Q = I$.

(b) For any power of A in the polynomial , say A^m, we have

$$(QAQ^T)^m = (QAQ^T)(QAQ^T)..(QAQ^T) = QA^m Q^T$$

since $Q^T Q = I$. The result follows.

(c) The result is true (by (b)) for polynomials, say $S_1(A)$ and $S_2(A)$. Yet the term-wise nature of the multiplication in the Hadarmard product implies that we must permute rows and columns of each factor similarly:

$$Q(S_1 \circ S_2)Q^T = (QS_1 Q^T) \circ (QS_2 Q^T).$$

The result follows.

(d) This is a sum of products of polynomials in A. So it follows from (c).

3.5 First show that $f_1 = f_2 = 0$ at the rest point (x_1^*, x_2^*). Then produce the linear system for small perturbations about this: for example, write $x_i(t) = x_i^* + y_i(t)$, substitute in and retain only the linear terms in the y_is. We have

$$\mathbf{df(x^*)} = \begin{pmatrix} -(p+q)^2 & -2p/(p+q) \\ (p+q)^2 & (p-q)/(p+q) \end{pmatrix}.$$

Thus determinant $\mathbf{df(x^*)} = (p+q)^2 > 0$, and trace $\mathbf{df(x^*)} = (p-q-(p+q)^3)/(p+q) < 0$.

Now we consider the spectrum of $\mathbf{df}(\mathbf{x}^*) - \mathrm{diag}(d_1, d_2)$. The result follows by examining when the determinant of this matrix become zero and hence there is a zero eigenvalue, and loss of stability as d_1 increases further. See also Grindrod (1995) [61].

3.6 Let $s = \lim_{n \to \infty}(1 - 1/n)^n$, then

$$\log s = \lim_{n \to \infty} (\log(1 - 1/n)/(1/n)).$$

Now we may use L'Hopital's rule

$$\log s = \lim_{n \to \infty} -1/(1 - 1/n) = -1,$$

so $s = e^{-1}$.

3.7

   (a) It is $1 - P(0|\infty, X) - P(1|\infty, X) = 1 - 2/e = 0.2642\ldots$

   (b) Each observation has a probability of $1/e$ of being resampled exactly one: so all n of them has a probability e^{-n}.

3.8 We have one hundred cases (H_1) and eighty controls (H_2), and an attribute called 'Gender' with values 0 and 1: sixty of the cases are 0s and forty are 1s, while thirty of the controls are 0s and fifty are 1s.

   Then we may use the models:

$$P(0|H_1, X) = 6/10, \ P(1|H_1, X) = 4/10, \ P(0|H_2, X) = 3/8, \ P(1|H_2, X) = 5/8.$$

So that

$$LBF(0) = \log(8/5), \ LBF(1) = \log(16/25).$$

Then,

$$ev(\text{Gender}) = \frac{60\log(8/5) + 40\log(16/25) - 30\log(8/5) - 50\log(16/25)}{180}$$

$$= \frac{30\log(8/5) - 10\log(16/25)}{180} = 0.103.$$

Next we have another attribute called 'county' with three values: Oxon, Berks, and Bucks. We know fifty of the cases are Oxon, thirty are Berks, and twenty are Bucks; while thirty of the controls are Oxon, thirty are Berks, and twenty are Bucks.

   Then we may use the models:

$$P(\text{Oxon}|H_1, X) = 5/10, \ P(\text{Berks}|H_1, X) = 3/10, \ P(\text{Bucks}|H_1, X) = 2/10,$$

$$P(\text{Oxon}|H_2, X) = 3/8, \ P(\text{Berks}|H_1, X) = 3/8, \ P(\text{Bucks}|H_2, X) = 2/8,$$

so that

$$LBF(\text{Oxon}) = \log(4/3), \ LBF(\text{Berks}) = \log(4/5), \ LBF(\text{Bucks}) = \log(4/5).$$

So
$$ev(\text{County})$$
$$= \frac{50 \log(1/3) + 30 \log(4/5) + 20 \log(4/5) - 30 \log(4/3) - 30 \log(4/5) - 20 \log(4/5)}{180}$$
$$= \frac{20 \log(4/3)}{180} = 0.032.$$

Hence the 'County' attribute is helpful but not nearly as useful as the 'gender' attribute in discriminating between cases and controls: on average it will not move the log odds as far in the correct direction.

Chapter 4

4.1 The relevant part of $\log \mathcal{L}$ in (4.3) is

$$\sum_{i=1}^{n} \sum_{k=1}^{K} \tau_{ik} \left(\log \pi_k \right) = \sum_{k=1}^{K} \log \pi_k \left(\sum_{i=1}^{n} \tau_{ik} \right).$$

Setting up the Lagrangian, with Lagrange multiplier μ for constraint $\sum_{k=1}^{K} \pi_k = 1$, then each π_k must maximise

$$\sum_{k=1}^{K} \log \pi_k \left(\sum_{i=1}^{n} \tau_{ik} \right) + \mu \left(1 - \sum_{k=1}^{K} \pi_k \right).$$

Thus

$$\frac{1}{\pi_k} \left(\sum_{i=1}^{n} \tau_{ik} \right) = \mu, \quad k = 1, \ldots, K.$$

Hence

$$\pi_k = \frac{1}{\mu} \sum_{i=1}^{n} \tau_{ik},$$

and the result follows since

$$1 = \sum_{k=1}^{K} \pi_k = \frac{1}{\mu} \sum_{k=1}^{K} \sum_{i=1}^{n} \tau_{ik} = \frac{n}{\mu}.$$

Chapter 5

5.1 Since $\mathbf{z} \in C_r$ then $P(D|H_j, X) = \hat{P}_{j,r}$ for some r for all j, so we have

$$O(D|H_k, X) = \frac{\hat{P}_{k,r}}{\sum_{j \neq k} \hat{P}_{j,r} \frac{P(H_j|X)}{(1-P(H_k|X))}} O(H_k|X).$$

If $N = 2$, then the middle ratio term simplifies since $P(H_j|X) = 1 - P(H_k|X)$ for $j \neq k$. Taking logs we have

$$\log O(D|H_k, X) = \log \frac{\hat{P}_{k,r}}{\hat{P}_{j,r}} + \log O(H_k|X)$$

where $l \neq k$. So take $k = 2$ and $j = 1$.

5.2 Using Stirling's approximation for the $\log(r_A!)$ term, then the contribution to $\log P(H_k|E)$ is

$$r_A \log(r_A \pm 1) - (r_A \pm 1) - r_A \log r_A + r_A - \frac{1}{2} \log 2\pi r_A + O(1/r),$$

$$= r_A (\log(r_A \pm 1) - \log r_A) \mp 1 - \frac{1}{2} \log 2\pi r_A + O(1/r)$$

$$= \pm 1 \mp 1 - \frac{1}{2} \log 2\pi r_A + O(1/r) = -\frac{1}{2} \log 2\pi r_A + O(1/r).$$

This is because for very large $r_k(A)$, the Poisson distribution gets very flat.

5.3 Using Stirling's approximation for the $\log(r_A!)$ term, then as before the contribution to $\log P(H_k|E)$ is

$$r_A \log(r_A \pm z) - (r_A \pm z) - r_A \log r_A + r_A - \frac{1}{2} \log 2\pi r_A + O(1/r),$$

$$= r_A (\log(r_A \pm z) - \log r_A) \mp z - \frac{1}{2} \log 2\pi r_A + O(1/r)$$

$$= \pm z \mp z - \frac{1}{2} \log 2\pi r_A + O(1/r) = -\frac{1}{2} \log 2\pi r_A + O(1/r).$$

5.4 Separation of variables implies:

$$\int_{1/2}^{f} \frac{df}{f(1-f)} = s.$$

Partial fractions

$$\int_{1/2}^{f} \frac{df}{f(1-f)} = \int_{1/2}^{f} \frac{1}{f} + \frac{1}{1-f} df = \log f/(1-f) = s.$$

So $f/(1-f) = e^s$, as required.

5.5 Directly we have

$$\frac{\partial}{\partial \beta_j} \log \mathcal{L} = \sum_{i=1}^{n} x_{i,j} y_i - \frac{f'(\beta^T x)}{(1 - f(\beta^T x))} x_{i,j}.$$

So substituting for f' we get

$$\frac{\partial}{\partial \beta_j} \log \mathcal{L} = \sum_{i=1}^{n} x_{i,j} y_i - \frac{f(\beta^T x)(1 - f(\beta^T x))}{(1 - f(\beta^T x))} x_{i,j}.$$

The result (5.6) follows.

Now starting from this last expression and differentiating it again with respect to $\beta_{j'}$, the only term inside the sum involving $\beta_{j'}$ is $f(\beta^T\mathbf{x})$. Using (5.4) we thus obtain

$$\frac{\partial^2}{\partial\beta_j\,\partial\beta_{j'}}\log\mathcal{L} = \sum_{i=1}^{n} -f(1-f)x_{i,j}x_{i,j'}$$

where f is evaluated at $\beta^T\mathbf{x}$, as required.

Chapter 6

6.1 Let week one be the first week of June in any given year. We need a suitable function $s^*(t|\lambda)$ that parametrizes these twin peaks to occur around weeks sixteen through nineteen and weeks twenty-nine through thirty-two, respectively.

6.2 Take the log of $P((A,\mathbf{y})|(\varepsilon,\sigma),X)$ and keep only the terms in the unknowns, σ and ε. We have

$$\left(\frac{-(\mathbf{y}-Ac)^T(\mathbf{y}-A\varepsilon)}{2\sigma^2}\right) - N\log\sigma.$$

So differentiate this with respect to ε and set the result equal to zero. We obtain the normal equation

$$0 = -A^T A\varepsilon + A^T\mathbf{y}.$$

This leads to the desired expression, provided $(A^T A)^{-1}$ exists.

Now differentiate with respect to σ and set the result equal to zero to obtain the formula for σ^2.

The handy part of this is that the equations decouple: we can solve for ε and then for σ. For non-Gaussian distributions, or when using non-conjugate priors, this is not the case and we expect to solve equations for the unknowns, σ and ε, simultaneously.

6.3 This a vector version of 'completing the square'.

Taking the log of $P((A,y)|(\varepsilon,\sigma),X).P_{\varepsilon\ \text{prior}}(\varepsilon|X)$ and considering this as a function of ε, we have

$$-\frac{(\mathbf{y}^T-\varepsilon^T A^T)(\mathbf{y}-A\varepsilon)}{2\sigma^2} - (\varepsilon^T-\varepsilon_0^T)C^{-1}\varepsilon - \varepsilon_0)/2 + \text{constant}$$

$$= -\varepsilon^T\left(\frac{C^{-1}}{2}+\frac{A^T A}{2\sigma^2}\right)\varepsilon + \varepsilon^T\left(\frac{A^T\mathbf{y}}{\sigma^2}+C^{-1}\varepsilon_0\right) + \text{constant},$$

$$= -\frac{1}{2}\varepsilon^T V\varepsilon + \varepsilon^T\left(\frac{A^T\mathbf{y}}{\sigma^2}+C^{-1}\varepsilon_0\right) + \text{constant},$$

where

$$V = C^{-1} + \frac{A^T A}{\sigma^2}.$$

Now for any ε^*, we have

$$-\frac{1}{2}(\varepsilon^T - \varepsilon^{*T})V(\varepsilon - \varepsilon^*) = -\frac{1}{2}\varepsilon^T V\varepsilon + \varepsilon^T V\varepsilon^* + \text{constant}.$$

So comparing these two quadratic forms, set

$$\varepsilon^* = V^{-1}\left(\frac{A^T\mathbf{y}}{\sigma^2} + C^{-1}\varepsilon_0\right).$$

Hence (6.6) ensures (6.5) as required.

6.4 There are obvious problems with Christmas weeks; we could consider leaving these points out of the data set. The underlying trend resembles the relationship $x_2 = At^m/(1 + Bt^m)$ where $x_1 = t$.

Chapter 7

7.1 For any i_{k+2} and i_k fixed, we have

$$P(i_{k+2}|i_k, X) = \sum_{j=1}^n P(i_{k+2}|j, X)P(j|i_k, X) = \sum_{j=1}^n A_{i_k j}A_{ji_{k+2}} = (A^2)_{i_k i_{k+2}},$$

where $A_{ij} = P(j|i, X)$ $i, j = 1, \ldots, n$, as before.
  So $\langle \mathbf{x}_{k+2}|\mathbf{x}_k, X\rangle = (A^2)^T\mathbf{x}_k = (A^T)^2\mathbf{x}_k$.

7.2 A is a non-negative irreducible matrix. So by the Perron–Frobenius theorem (see Chapter 1) it has a spectral radius of one, since all rows of A must sum to one. The Perron–Frobenius right eigenvector of A is $\mathbf{s} = (1, 1, \ldots, 1)^T$. Let $\mathbf{e}^*$ be the positive normalized left eigenvector of A (the right eigenvector of A^T) such that $\mathbf{s}^T\mathbf{e} = 1$, corresponding to the Perron–Frobenius eigenvalue (which is simple). Then we have $A^T\mathbf{e}^* = \mathbf{e}^*$. Also the Perron–Frobenius right eigenvector of A is $\mathbf{s}$.
  Then for any $\mathbf{x}_k$ conservation of mass applies:

$$\mathbf{s}^T\mathbf{x}_{k+1} = \mathbf{s}^T(A^T\mathbf{x}_k) = (\mathbf{s}^T A^T)\mathbf{x}_k = \mathbf{s}^T\mathbf{x}_k.$$

Thus the sequence $\{\mathbf{x}_k\}$ remains on the simplex $\{\mathbf{x} \geq 0 | \mathbf{s}^T\mathbf{x}_k = \text{constant}\}$.
  As $k \to \infty$ we thus have $(A^T)^k\mathbf{x}_1 \to (\mathbf{s}^T x\mathbf{x}_1)\mathbf{e}$ since the spectral radius of A^T on the subspace orthogonal to $\mathbf{e}$ is strictly less than one.

7.3 Assuming that the transitions are independent then likelihood of the data given the model, $\mathcal{L} = P(\text{Data}|\text{Model}, X)$, is obtained by multiplying together the likelihoods of each transition. We observed the ij-transition exactly m_{ij} times and it has a probability equal to A_{ij}. Thus

$$\mathcal{L} = \prod_{i=1,\,j=1}^{n} A_{ij}^{M_{ij}}.$$

Substituting for A_{ij} and taking logs we obtain

$$\log \mathcal{L} = \sum_{i=1,\,j=1}^{n} M_{ij}(\log(1 + M_{ij}) - \log(n + r_i)).$$

Hence, rearranging the last term

$$\log \mathcal{L} = \sum_{i=1,\,j=1}^{n} M_{ij} \log(1 + M_{ij}) - \sum_{i=1}^{n} r_i \log(n + r_i).$$

In fact it is not a good idea to vary the model (by optimizing state definitions) so as to maximize $\log \mathcal{L}$. Suppose we simply assign all sequential observations to the same state, state one. Then there is only one transition that is ever observed: $1 \rightarrow 1$ that has a probability almost equal to one. So the likelihood of the observed state sequence is almost one. It cannot get any better likelihood wise, but could not be any worse in terms of the usefulness of the information conveyed.

REFERENCES

[1] Aczel, J. (1996), *Lectures on Functional Equations and Their Applications*, Academic Press: New York, 1966; and Dover, New York; 2006.

[2] Allen M.R. and Smith L.A. (1996), 'Monte Carlo SSA: Detecting Irregular Oscillations in the Presence of Colored Noise', *Journal of Climate*, 9 (12): 3373–3404.

[3] Alter O., Brown P. O., Botstein D. (2003), '*Generalized Singular Value Decomposition for Comparative Analysis of Genome-Scale Expression Data Sets of Two Different Organisms*', *Proceedings of the National Academy of Sciences*, 100: 3351–3356.

[4] Aral S. (2012), '*Social science: Poked to vote*', *Nature*, 489: 212–214.

[5] Aral S. and Walker D. (2012), 'Identifying influential and susceptible members of social networks', *Science*, 337: 337–341.

[6] Ariely, D. (2008), *Predictably Irrational: The Hidden Forces That Shape Our Decisions*, Harper-Collins:

[7] Backstrom L., Boldi P., Rosa M. and Ugander J. (2012), In: 'Four degrees of separation', *WebSci '12 Proceedings of the 3rd Annual ACM Web Science Conference*, pp. 33–42 ACM: New York.

[8] Bader J. S., Chaudhuri A., Rothberg J. M., Chant J. (2003), 'Gaining confidence in high-throughput protein interaction networks', *Nature Biotechnology* 22: 78–85.

[9] Bakshy E., Hofman J. M., Mason W. A., and Watts D. J. (2011), 'Everyone's an influencer: quantifying influence on Twitter'. In: *Proceedings of the 4th ACM International Conference on Web Search and Data Mining*, WSDM '11, pp. 65–74, ACM: New York.

[10] Bakshy E., Rosenn I., Marlow C., and Adamic L. (2012), 'The Role of Social Networks in Information Diffusion'. In: *Proceedings of the 21st International Conference on World Wide Web*, WWW '12, pp. 519–528 ACM: New York.

[11] Barabsi A.-L., and Albert R. (1999), 'Emergence of Scaling in Random Networks', *Science* 286 (5439): 509–512.

[12] Barrat A., Barthlemy M., Vespignani A. (2008), *Dynamical Processes on Complex Networks*, Cambridge: Cambridge University Press.

[13] Bollobás, B. (2013), *Modern Graph Theory (Graduate Texts in Mathematics)* (Book 184), Springer-Verlag: New York; Corrected edition.

[14] Bollobas B., Riordan O., Spencer J., and Tusnady G. (2001), 'The Degree Sequence of a Scale-Free Random Graph Process', *Random Structures and Algorithms* 18 (3): 279–290.

[15] Bonchi F., Castillo C., Gionis A., and Jaimes A. (2011), 'Social Network Analysis and Mining for Business Applications', *A Transactions on Intelligent Systems and Technology*, 2(3): 22.

[16] Borge-Holthoefer J., Baos R. A., Gonzalez-Bailon S., and Moreno Y. (2013), 'Cascading Behaviour in Complex Socio-Technical Networks', *Journal of Complex Networks*, 1: 324.

[17] Brazma A., Parkinson H., Schlitt T., and Shojatalab, M. (2001), 'A Quick Introduction to Elements of Biology - Cells, Molecules, Genes, Functional Genomics, Microarrays', *European Bioinformatics Institute*, online tutorial.

[18] Broomhead D.S. and King G.P. (1986), 'Extracting Qualitative Dynamics from Experimental Data', *Physica D*, 20: 217–236.

[19] Carlsson G. (2009), 'Topology and Data', *Bulletin (New Series) of the AMS*, Volume 46, Number 2: 255–308.

[20] Centola D. (2010), 'The spread of Behavior in an Online Social Network Experiment', *Science*, 329, p. 1194.

[21] Cha M., Haddadi H., Benevenuto F., and Gummadi K. P. (2010), 'Measuring User Influence in Twitter: The Million Follower Fallacy', In: *ICWSM 10: Proceedings of International AAAI Conference on Weblogs and Social*.

[22] Charlton N. A., Greetham D. V., Grindrod P., Haben S. A., and Singleton C. P. (2013), A Probabilistic Framework for Forecasting Spiky Demand profiles, November 2013, preprint: http://ora.ox.ac.uk/objects/uuid:68b6552b-4b96-4384-a7c6-9c8cd55b6c8d

[23] Charlton N.A., Greetham D. V., and Singleton C.P. (2014), 'Graph-based algorithms for comparison and prediction of household-level energy use profiles'. To appear in: *Proceedings of IWIES 2013 (International Workshop on Intelligent Energy Systems)*.

[24] Choi J.K., Yu U., Yoo O.J., and Kim S. (2005), 'Differential Coexpression Analysis Using Microarray Data and Its Application to Human Cancer', *Bioinformatics* 21: 4348–4355.

[25] Chow S.N. and Hale J.K. (1982), *Methods Of Bifurcation Theory*. Springer-Verlag: New York.

[26] Cignifi Inc (2011), 'Building the Bridge to New Consumers in Brazil', White Paper, <http://www.cignifi.com/en-us/technology/case-study>, Cignifi Inc.

[27] Ciulla F., Mocanu D., Baronchelli A., Goncalves B., Perra N., and Vespignani A. (2012), 'Beating the News Using Social Media: The Case Study of American Idol', *EPJ Data Science*, 1: 111.

[28] Clarke S. and Grindrod P. (2007), 'A Bayesian Estimation of Unfolded Pitchfork Bifurcation Structure Based Upon Experimental Data', *IMA Journal of Applied Mathematics*, 72: 395–404.

[29] Cohen J. (2004), 'Bioinformatics—An Introduction for Computer Scientists', *ACM Computing Surveys*, 36: 122–158.

[30] Coombes K.R., et al. (2005). Improved Peak Detection and Quantification of MS Data Acquired from Surface-Enhanced Laser Desorption and Ionization by Denoising Spectra with the Undecimated Discrete Wavelet Transform. *Proteomics* 5(16): 4107–4117.

[31] Cortes C. and Vapnik V.N. (1995), 'Support-Vector Networks', *Machine Learning*, 20(3): 273–297 <http://www.springerlink.com/content/k238jx04hm87j80g/>

[32] D'Agostini G. (2003), *Bayesian Reasoning in Data Analysis—A Critical Introduction*, World Scientific Publishing, Singapore.

[33] Davenport T.H. and Harris J.G. (2007), *Competing on Analytics: The New Science of Winning*, Harvard Business School Press, pp. 240.

[34] Delrieu O. and Bowman C.E. (2006), 'Visualizing Gene Determinants of Disease in Drug Discovery', *Pharmacogenomics* 7(3): 311–329.

[35] Dewdney A.K. (1997), *Yes, We Have No Neutrons: An Eye-Opening Tour through the Twists and Turns of Bad Science*, NY: Wiley.

[36] Do K-A. (2006), *Bayesian Inference for Gene Expression and Proteomics*, Cambridge: Cambridge University Press.

[37] Eagle N., Pentland A., and Lazer D. (2009), 'Inferring Social Network Structure Using Mobile Phone Data', *Proceedings of the National Academy of Sciences*, 106(36): 15,274–15,278.

[38] Efron B. and Tibshirani R. (1993), *An Introduction to the Bootstrap*. Boca Raton, FL: Chapman & Hall/CRC.

[39] Elliot W.H. and Elliot D.C. (2003), *Biochemistry and Molecular Biology*, Second Edition, Oxford: Oxford University Press.

[40] Estrada E. and Hatano N. (2008), 'Communicability in Complex Networks'. *Physical Review* E 77, 036111.

[41] Farhi P. (2003), Oreos Tweeted ad was Super Bowl Blackout's Big Winner, Washington Post, (February 05, 2013).

[42] Feigenbaum M.J. (1978), Quantitative Universality for a Class of Non-Linear Transformations, *Journal of Statistical Physics* 19: 25–52.

[43] Feller W. (1968), *Introduction to Probability Theory and its Applications*, Volume 1, third edition. New York: Wiley.

[44] Filkov V., Skiena S., and Zhi J. (2002), 'Analysis Techniques for Microarray Time-Series Data', *Journal of Computational Biology* 9: 317–330.

[45] Fogel P., Young S.S., Hawkins D.M., and Ledirac N. (2007), 'Inferential, Robust Non-Negative Matrix Factorization Analysis of Microarray Data', *Bioinformatics* 23: 44–49.

[46] Friston K.J. and Stephan K.E. (2007), 'Free Energy and the Brain', *Synthese*, 159: 417–456.

[47] Gleave E., Welser H.T., Lento T. M., and Smith M. A. (2009), 'A Conceptual and Operational Definition of Social Role in Online Community'. In: *Proceedings of the 42nd Hawaii International Conference on System Sciences*, Los Alamitos, CA, USA, IEEE Computer Society: pp. 111.

[48] Goldberg D.E. (1989), *Genetic Algorithms in Search, Optimization and Machine Learning*, Kluwer Academic Publishers: Boston, MA.

[49] Golub G.H. and Van Loan, C.F. (1996), *Matrix Computations*, third edition, The Johns Hopkins University Press: London.

[50] Golub T. R., Slonim D. K., Tamayo P., Huard C., Gaasenbeek M., Mesirov J. P., Coller H., Loh M. L., Downing J. R., Caliguri M. A., Bloomfield C. D. and Lander E. S. (1999), 'Molecular Classification of Cancer: Class Discovery and Class Prediction by Gene Expression Monitoring', *Science* 286: 531–537.

[51] Golubitsky, M. and Schaeffer D. G. (1985), *Singularities and Groups in Bifurcation Theory* Volume 1, Springer-Verlag: New York.

[52] Grindrod P. (2002), 'Range-Dependent Random Graphs and Their Application to Modeling Large Small-World Proteome Datasets', *Physics Review* E 66: 066702.

[53] Grindrod P. and Kibble M. (2004), 'Review of Uses of Network and Graph Theory Concepts Within Proteomics', *Expert Review of Proteomics* 1: 229–238.

[54] Grindrod P., Higham, D. J., Vass J.K., and Spence A. (2006), 'Systems Biology: Unravelling Complex Networks', *University of Strathclyde Mathematics Research Report* 17.

[55] Grindrod P., Higham D.J., Kalna G., Spence A., Stoyanov Z., and Vass J.K. (2008), 'DNA Meets the SVD', *Mathematics Today*, 44: 80–85.

[56] Grindrod P. and Higham D.J. (2013), 'A Matrix Iteration for Dynamic Network Summaries', *SIAM Review*, 55: 118–128.

[57] Grindrod P. and Higham D.J. (2010), 'A Dynamical Systems View of Network Centrality', *Proceedings of the Royal Society* A, submitted.

[58] Grindrod P. and Waters, D.J. (2003), 'Customer Centric Strategy: Success or Survival?', *Executive Outlook*, 3(4) (December 2003), 74–81.

[59] Grindrod P., Higham D.J., Parsons M.C., and Estrada E. (2011), Communicability Across Evolving Networks. *Physical Review* E 83 2011, 046120.

[60] Grindrod P., Higham D.J. and Parsons, M.C. (2012), Bistability Through Triadic Closure. Internet Mathematics 8: 402–423.

[61] Grindrod P. (1995), *Patterns and Waves*, 2nd Edition, Oxford: Oxford University Press.

[62] Grindrod P. and Parsons M.C. (2012), Complex Dynamics in a Model of Social Norms. Draft submitted. Preprint MPS-2012-21, Department of Mathematics and Statistics, University of Reading, https://www.reading.ac.uk/web/FILES/maths/Preprint_12_21_Parsons.pdf.

[63] Grindrod P., Stoyanov Z.V., Smith G.M., and Saddy J.D. (2013), 'Primary Evolving Networks and the Comparative Analysis of Robust and Fragile Structures', *Journal of Complex Networks* Volume 2, Issue 1: 60–73.

[64] Grindrod P., Stoyanov Z.V., Smith G.M., and Saddy J.D. (2014), Resonant Behaviour of Irreducible Networks of Neurons with Delayed Coupling, submitted *Physica D*, 2014.

[65] Gronroos C. (2000), Sevice, Management and Marketing, A Customer Relationship Management Approach, Wiley: Chichester.

[66] Gupta S., Hanssens D., Hardie B., Kahn W., Kumar V., Lin N. and Sriram S. (2006), 'Modeling Customer Lifetime Value', *Journal of Service Research*, 9(2): 139–155.

[67] Gupta S., Lehmann D.R., and Stuart J.A. (2004), 'Valuing Customers', *Journal of Marketing Research*, 41(1): 7–18.

[68] Haben S.A., Ward J.A., Greetham D.V., Grindrod P., and Singleton C.P. (2013), 'A New Error Measure for Forecasts of Household-level, High Resolution Electrical Energy Consumption'. To appear in. *The International Journal of Forecasting*, 30: 246–256, 2014.

[69] Hand D.J. (2006), Classifier Technology and the Illusion of Progress, Statistical Science 21(1): 1–112.

[70] Henry D. (1981), *Geometrical Theory of Semilinear Parabolic Equations*, Lecture Notes in Math 840, Springer-Verlag: New York.

[71] Hardiman G. (2004), 'Microarray platforms-Comparisons and Contrasts', *Pharmacogenomics* 5: 487–502.

[72] Heath R. (2012), *Seducing the Subconscious: The Psychology of Emotional Influence in Advertising*, Wiley-Blackwell.

[73] Higham D. J. (2005), 'Spectral Reordering of a Range-Dependent Weighted Random Graph. The Institute of Mathematics and Its Applications, *(IMA)' Journal of Numerical Analysis*, 25: 443–457.

[74] Higham D.J., Kalna G., and Vass J.K. (2005), 'Analysis of the Singular Value Decomposition as a Tool for Processing Microarray Expression Data. In: Proceedings of ALGORITMY 2005, Podbanské, Slovakia.

[75] Higham D.J., Kalna G., and Kibble M. (2007), 'Spectral Clustering and Its Use in Bioinformatics', *Journal of Computational and Applied Math* 204: 25–37.

[76] Higham D.J., Kalna G., and Vass J.K. (2007), Spectral Analysis of Two-Signed Microarray Expression Data, *IMA Mathematical Medicine and Biology*, 24: 131–148.

[77] Higham D.J., Grindrod P., Mantzaris A.V., Ainley F., Otley A., and Laflin P. (2014), 'Anticipating Activity in Social Media Spikess', *Proceedings of the National Academy of Science*, submitted.

[78] Higham N.J. (2008), 'Functions of Matrices: Theory and Computation', *Society for Industrial and Applied Mathematics*, Philadelphia, PA, USA.

[79] Horn R.A. and Johnson C.R. (1985), *Matrix Analysis*, Cambridge University Press.

[80] Hu H. and Wanga X. (2009), 'Evolution of a Large Online Social Network', *Physics Letters A*, 12–13: 1105–1110.

[81] Huffaker D. (2010), 'Dimensions of Leadership and Social Influence in Online Communities', *Human Communication Research*, 36: 593–617.

[82] Irish National Smart Meter Trial Data <http://www.seai.ie/News_Events/Press_Releases/2012/Full_Data_from_National_Smart_Meter_Trial_Published.html>

[83] Jaynes E.T. and Bretthorst, G.L. (2003), *Probability Theory: The Logic of Science*. Cambridge University Press: Cambridge.

[84] Kahneman D. (2011), *Thinking, Fast and Slow*, New York: Farrar, Straus and Giroux.

[85] Kanehisa M. (2003), *Post-genome Informatics*, Oxford: Oxford University Press.

[86] Katz L. (1953), A Newstatns Index Derived from Sociometric Analysis, Psychometrika, 18(1): 39–43.

[87] Knight P. A. (2006), 'The Sinkhorn-Knopp Algorithm: Convergence and Applications'. Technical Report TR/PA/06/42, *The Parallel Algorithms Project*, CERFACS, France.

[88] Kolda T.G. and Bader B.W. (2009), 'Tensor Decompositions and Applications'. *SIAM Review* 51: 455–500.

[89] Kuczma M., Choczewski B., and Ger R. (1990), 'Iterative Functional Equations'. *Encyclopedia of Mathematics and its Applications*, Vol. 32, Cambridge University Press: Cambridge.

[90] Kwak H., Lee C., Park H., and Moon S. (2010), 'What is Twitter, a Social Network or a News Media? In: Proceedings of the 19th International Conference on World Wide Web, WWW '10, New York, NY, USA, ACM, 591–600.

[91] Laflin P., Mantzaris A. V., Ainley F., Otley A., Grindrod P., and Higham D. J. (2013), 'Discovering and Validating Influence in a Dynamic Online Social Network', *Social Network Analysis and Mining*, 3: 1311–1323.

[92] Lazer D., Pentland A., Adamic L., Aral S., Barabasi A.-L., Brewer D., Christakis N., Contractor N., Fowler J., Gutmann M., and Jebara T. (2009), 'Computational Social Science', *Science*, 323: 721–723.

[93] Lerman K., Ghosh R., and Surachawala T. (2012), 'Social Contagion: An Empirical Study of Information Spread on Digg and Twitter Follower Graphs', arXiv: 1202.3162.

[94] Leskovec J., Adamic L.A., and Huberman B.A. (2007), The Dynamics of Viral Marketing, ACM Transaction on the Web, 1.

[95] Lehmann J., Goncalves B., Ramasco J.J., and Cattuto C. (2012), 'Dynamical Classes of Collective Attention in Twitter'. In: Proceedings of the 21st International Conference on World Wide Web, WWW '12, New York, NY, USA, ACM: 251–260.

[96] Li J., Li X., Su H., Chen H., Galbraith D. W. (2006), 'A Framework of Integrating Gene Relations from Heterogeneous Data Sources: An Experiment on Arabidopsis Thaliana', *Bioinformatics* 22: 2037–2043.

[97] Lin Y.-R., Keegan B., Margolin D., and Lazer D. (2013), 'Rising Tides or Rising Stars?: Dynamics of Shared Attention on Twitter During Media Events', arXiv: 1307.2785 [cs.SI].

[98] Lo K. and Gottardo R. (2007), 'Flexible Empirical Bayes Models for Differential Gene Expression', *Bioinformatics* 23: 328–335.

[99] Mantzaris A. V. and Higham D. J. (2012), A Model for Dynamic Communicators, *European Journal of Applied Mathematics*, 23: 659–668.

[100] Mantini D. (2007), 'LIMPIC: A Computational Method for the Separation of Protein MALDI-TOF Signal From Noise, *BMC Bioinformatics*, 8: 101. <www.biomedcentral/147-2105-8-101>

[101] McLachlan G. and Peel D. (2000), *Finite Mixture Models*, Wiley-Blackwell.

[102] McCullagh P. and Nelder J.A. (1989), *Generalized Linear Models*, Second Edition CRC Monographs on Statistics & Applied Probability, Chapman and Hall.

[103] Metropolis, N. et al. (1953), 'Equations of State Calculations by Fast Computing Machines', *Journal of Chemical Physics*, 21: 1087–1091.

[104] Milgram S. (1967), 'The Small World Problem', *Psychology Today*, Vol. 2: 60–67.

[105] Morrison J. L., Breitling R., Higham D. J., and Gilbert D. R. (2006), 'A Lock-and-Key Model for Protein-Protein Interactions', *Bioinformatics* 22: 2012–2019.

[106] Mumford D. (1992), 'On The Computational Architecture of the Neocortex. II The Role of Cortico-Cortical Loops'. *Biological Cybernetics*, 66: 241–251.

[107] Murray J.D. (2002), 'Mathematical Biology, An Introduction', *Interdisciplinary Applied Mathematics*, Vol. 17, Springer.

[108] Newman M. E. J. (2010), *Networks. An Introduction*, Oxford University Press: Oxford.

[109] O'Hagan A. and Forster J. (2003), *Kendall's Advanced Theory Of Statistics, Volume 2B: Bayesian Inference*, Wiley–Blackwell.

[110] Pfeifer P.E. and Carraway R. L. (2000), 'Modeling Customer Relationships as Markov Chains', *Journal of Interactive Marketing*, 14(2): 43–55.

[111] Press W.H., Teukolsky S.A., Vetterling W.T., and Flannery B.P. (2007), *Numerical Recipes: The Art of Scientific Computing*, third edition, New York: Cambridge University Press.

[112] Romero D. M., Meeder B., and Kleinberg J. (2011), 'Differences in the Mechanics of Information Diffusion Across Topics: Idioms, Political Hashtags, and Complex Contagion on Twitter'. In: *Proceedings of the 20th International Conference on World Wide Web*, WWW '11, New York, NY, USA, ACM: 695–704.

[113] Shafiq M. Z. and Liu A. X. (2013), Modeling morphology of social network cascades, CoRR, abs/1302.2376.

[114] de Silva E., Stumpf M. P. H. (2005), 'Complex Networks and Simple Models in Biology', *Royal Society Interface* 2: 419–430.

[115] Shamma D. A., Kennedy L., and Churchill E. F. (2011), 'In the Limelight Over Time: Temporalities of Network Centrality', *Proceedings of the 29th International Conference on Human Factors in Computing Systems* CSCW 2011, ACM.

[116] Schulz-Trieglaff O. (2007), 'A Fast and Accurate Algorithm for the Quantification of Peptides from MS Data', *Proceedings of the 11th Annual International Conference on Resvised on Computational Molecular Biology*, (RECOM 2007).

[117] Spence A., Stoyanov Z., and Vass J. K. (2007), 'The Sensitivity of Spectral Clustering Applied to Gene Expression Data', *1st International Conference on Bioinformatics and Biomedical Engineering*, ICBBE 2007.

[118] Strang G. (1989), 'Wavelets and Dilation Equations: A Brief Introduction', *SIAM Review*, 31(4): 614–627.

[119] Tang L.L., Thomas L.C., Thomas S. and Bozzetto J-F. (2005), 'Modelling the Purchase Dynamics of Insurance Customers Using Markov Chains', Discussion Papers in *Centre for Operational Research, Management Science and*

Information Systems, CORMSIS-05-02. University of Southampton. <http://www.management.soton.ac.uk/research/publications/documents/CORMSIS-05-02.pdf>

[120] Titz B., Schlesner M., and Uetz P. (2004), 'What Do We Learn from High-Throughput Protein Interaction Data?, *Expert Review of Proteomics* 1: 111–121.

[121] Uetz P., Giot L., Cgney G., Mansfield T. A., Judson R. S., Knight J. R., Lockshon E., Narayan V., Srinivasan M., Pochart P., Qureshi-Emili A., Li Y., Godwin B., Conover D., Kalbfleish T., Vijayadamodar G., Yang M., Johnston M., Fields S., and Rothberg J. M. (2000), 'A Comprehensive Analysis of Protein-Protein Interactions in Sacharomyces Cerevisiae', *Nature* 403: 623–627.

[122] Van Fleet P. (2008), *Discrete Wavelet Transformations: An Elementary Approach with Applications*, city/country Wiley:

[123] Ward J.A. and Grindrod P. (2013), 'Aperiodic dynamics in a deterministic model of attitude formation in social groups, arXiv 1302.0164'.

[124] Watts D.J. and Strogatz S.H. (1998), 'Collective dynamics of "small-world" networks'. *Nature* 393 (6684): 440–442.

[125] Watts D.J. (1999), *Small Worlds: The Dynamics of Networks between Order and Randomness*, Princeton University Press.

[126] Xiaoli L., Jin L., and Xin Y. (2007), 'A Wavelet-Based Data Pre-Processing Analysis Approach in Mass Spectrometry', *Computers in Biology and Medicine* 37: 509–516. <http://www.cs.bham.ac.uk/ xin/papers/CompInBioMed2007.pdf>

INDEX